W9-BGA-302

APOLLO

ZACK SCOTT

ABRAMS IMAGE, NEW YORK

INTRODUCTION

The Apollo program, which ran from 1961 until 1972, will be forever remembered as a milestone in human endeavour. It was an extraordinary accomplishment, that demanded huge technological leaps, a colossal amount of funding, and a sizeable, highly skilled workforce. Project Apollo was the largest commitment of resources made by any nation during peacetime; at its height it employed more than 400,000 people and cost $24 billion in total, more than $110 billion in today's money.

The reason the United States devoted such resources to the program was because it was engaged in the Space Race with the Soviet Union. In the aftermath of the Second World War, deep political and economic differences created a rivalry between the two superpowers, which led to the Cold War. Although there was no full-scale armed combat directly between the two sides, each country strove for economic, scientific, and military superiority. By mastering the techniques necessary for space travel, each side was not only showing the other how much more advanced they were, but that they potentially had the means to deploy a nuclear bomb anywhere in the world. Having launched the first satellite (Sputnik I) in 1957, the Soviets would again beat the United States by sending the first man into space, cosmonaut Yuri Gagarin, on April 12, 1961. America was clearly lagging behind. In an effort to overtake the Soviets, President John F. Kennedy laid down the challenge of "landing a man on the Moon and returning him safely to the Earth." And so the Apollo program was born.

Apollo was NASA's third human spaceflight program. The first, Project Mercury, began in 1958 and ran until 1963. Its primary goal was to place a manned craft in Earth orbit, which it did on four occasions. Once NASA had proved they could send a man into space, they then commenced with Project Gemini, which would run alongside the Apollo program from 1961 until 1966. Its objectives were to test the space travel techniques that would be necessary on the Apollo program.

The Mercury and Gemini programs had prepared NASA's scientists, engineers, and astronauts as far as possible, but with Apollo there were many more challenges yet to face. Through the determination, focus, and coordinated efforts of the thousands who supported the missions, the Apollo program would become the pinnacle of human achievement, and forever a testament to what the human race can achieve when it sets its goals high enough.

WE CHOOSE TO GO TO THE

MOON

MACHINERY

To achieve the goal of sending someone to the Moon and back, the scientists at NASA decided on a technique they called Lunar Orbit Rendezvous. This meant that they would send a spacecraft coupled with a landing craft to orbit the Moon. Once in the Moon's orbit the landing craft would detach and carry its passengers down to the surface, where they could then explore. On their return, the astronauts would lift off from the Moon in a portion of their landing vehicle and return to the orbiting spacecraft. After they transferred back into the main spacecraft, the remaining section of their landing vehicle would then be discarded, and they would travel back to Earth. The main spacecraft was known as the Command/Service Module, and the landing craft was called the Lunar Module.

These spacecraft would not have the ability to get to the Moon on their own. In order to escape the pull of Earth's gravity, a huge rocket was needed, so the Saturn V was created. It was a three-stage rocket, meaning that it was made from three parts that would fire one after the other, with each part being dispensed with after use. The "V" in Saturn V refers to the five huge F1 engines that blasted the rocket skyward at liftoff.

Although the Saturn V would become famous for getting man to the Moon, smaller rockets were also used during the Apollo program. Early unmanned missions used Little Joe II, Saturn I, and the Saturn IB to test the rocket and guidance technology, as well as to take measurements and readings in preparation for manned missions. Aside from the Saturn V, the Saturn IB was the only other rocket from the program to take part in a manned mission, which it did on just one occasion.

LAUNCH ESCAPE SYSTEM

This system was connected to the Command Module (CM) of crewed spacecraft and was powered by solid-fuel rockets. Its purpose was to quickly separate the CM from the rest of the rocket in case of emergency, specifically situations where there was an imminent threat to the crew, such as an impending explosion. It could be used up to an altitude of 30 km.

INSTRUMENT UNIT

This was situated above the Saturn V's third-stage rocket. It was a ring of instruments that provided the guidance for the Saturn rockets. Among its components were computers, control electronics, accelerometers, and gyros. After it was no longer needed, it was jettisoned from the Apollo spacecraft and sent into either Earth or solar orbit, or to crash into the Moon.

FINS

On each of the Saturn rockets, fins surrounded the engines at the base of the first-stage rocket. They were there to provide aerodynamic stability.

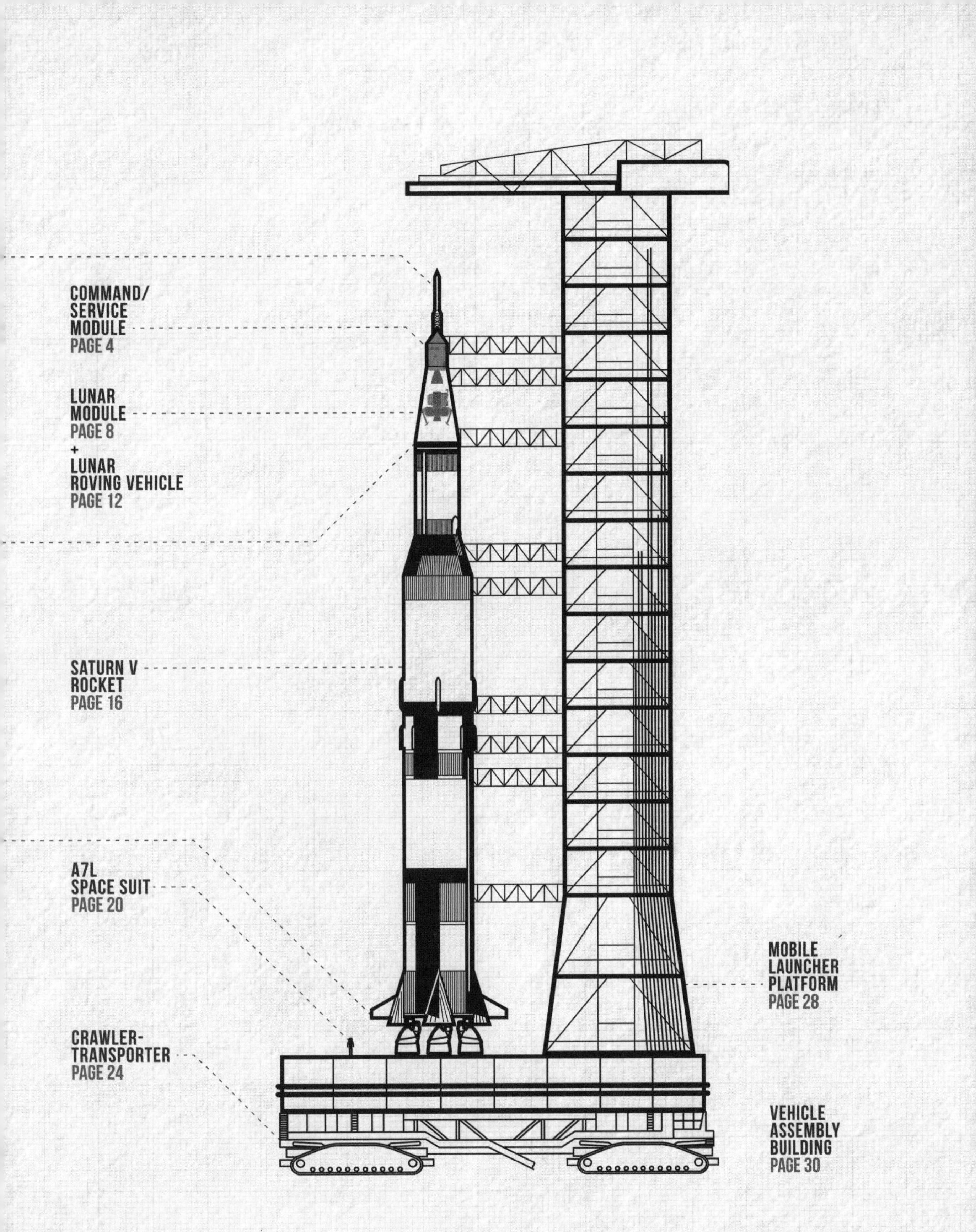
COMMAND/
SERVICE
MODULE
PAGE 4
LUNAR
MODULE
PAGE 8
+
LUNAR
ROVING VEHICLE
PAGE 12
SATURN V
ROCKET
PAGE 16
A7L
SPACE SUIT
PAGE 20
CRAWLER-
TRANSPORTER
PAGE 24
MOBILE
LAUNCHER
PLATFORM
PAGE 28
VEHICLE
ASSEMBLY
BUILDING
PAGE 30

COMMAND/SERVICE MODULE CSM

The term Command/Service Module (CSM) refers to two modules: the Command Module (CM) and the Service Module (SM). These modules remained attached to each other until the end of a mission.

The three crew members were based in the CM, which was pressurized with oxygen and nitrogen and maintained a comfortable temperature. They were stationed behind their instruments and controls in adjustable couches that provided a variety of different positions, depending on the phase of the mission and flight situation. The five windows on the CM allowed the crew to see out into space, and helped them when docking with the Lunar Module. The CM had twelve thrusters that were used after the craft had separated from the SM, controlling its reentry into the Earth's atmosphere.

PITCH THRUSTERS
Also known as reaction control jets. There were 12 thrusters on the CM, controlling pitch, roll, and yaw. They each produced 445 N of thrust, operating in bursts, from 12 milliseconds to 500 seconds.

DROGUE PARACHUTES
After reentering the Earth's atmosphere, drag would slow the craft to 480 km/h. The two drogues would then deploy, slowing the CM to 200 km/h.

DOCKING PROBE
Used to connect the CM to the LM, and to allow the crew to move between the two modules when docked.

MAIN PARACHUTES
These were released after the drogue parachutes, slowing the craft down to 35 km/h. Safe splashdown requires 2 of 3 to be successfully deployed.

COMMAND MODULE

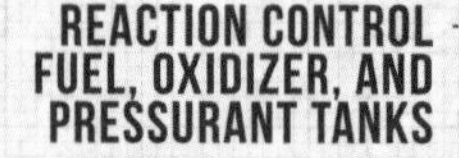

STORAGE COMPARTMENTS
Situated throughout the CM, they held scientific equipment and enough food to sustain 3 people for 11 days.

YAW THRUSTERS

SERVICE MODULE

The engine of the Service Propulsion System (SPS) was mounted at the rear of the SM. This provided the vehicle's main thrust once the Saturn V's rockets had been used and discarded. The SM's Reaction Control System (RCS) comprised four thruster quads that steered the spacecraft. Inside the SM were fuel and oxidizer tanks, as well as tanks for pressurant, which was needed to push the propellants through the engine. The SM also housed fuel cells and batteries that provided electrical power to the CM. At the end of a mission, just before the craft would reenter the Earth's atmosphere, the SM was cast off and the CM would return the crew to Earth. Friction with the air would cause the SM to burn up as it reentered.

COMMAND/SERVICE MODULE CSM

Total Mass:	30,080 kg
Dry Mass:	11,165 kg

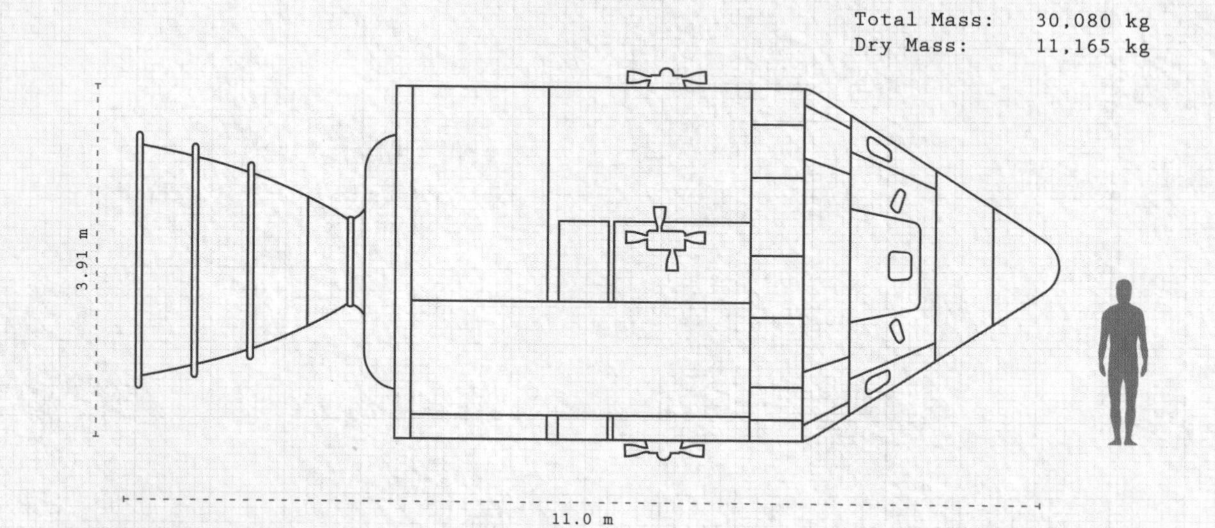

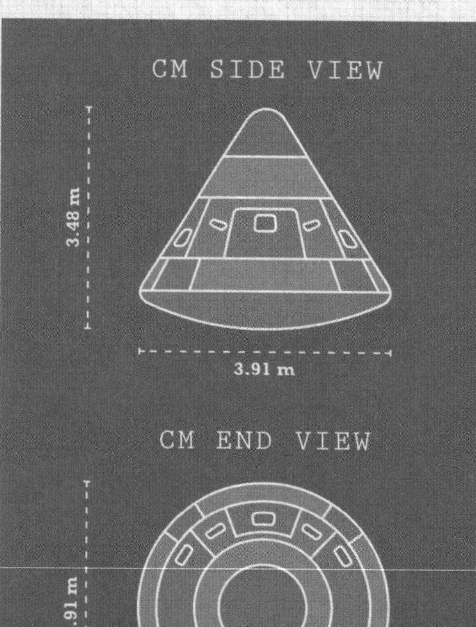

COMMAND MODULE CM

Crew Members:	3
Total Mass (inc. crew):	5,560 kg
Unloaded Mass:	5,150 kg
Habitable Volume:	5.90 m^3
Pressurized Volume:	7.65 m^3
Drinking Water Capacity:	15 kg
Waste Water Capacity:	26.5 kg
Batteries:	5
Total Battery Charge:	121.5 AH
Drogue Parachutes:	2 x 5 m
Pilot Parachutes:	3 x 2.2 m
Main Parachutes:	3 x 25.4 m

CM INTERIOR SPACE

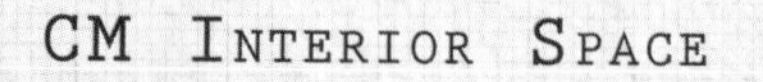

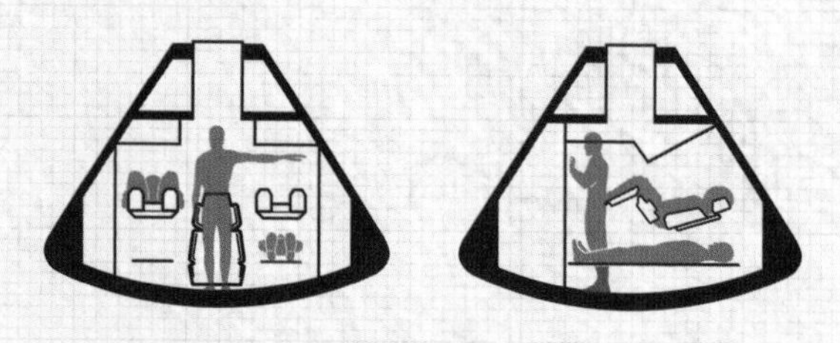

Service Module SM

Total Mass: 24,520 kg
Unloaded Mass: 6,015 kg

SM Tanks

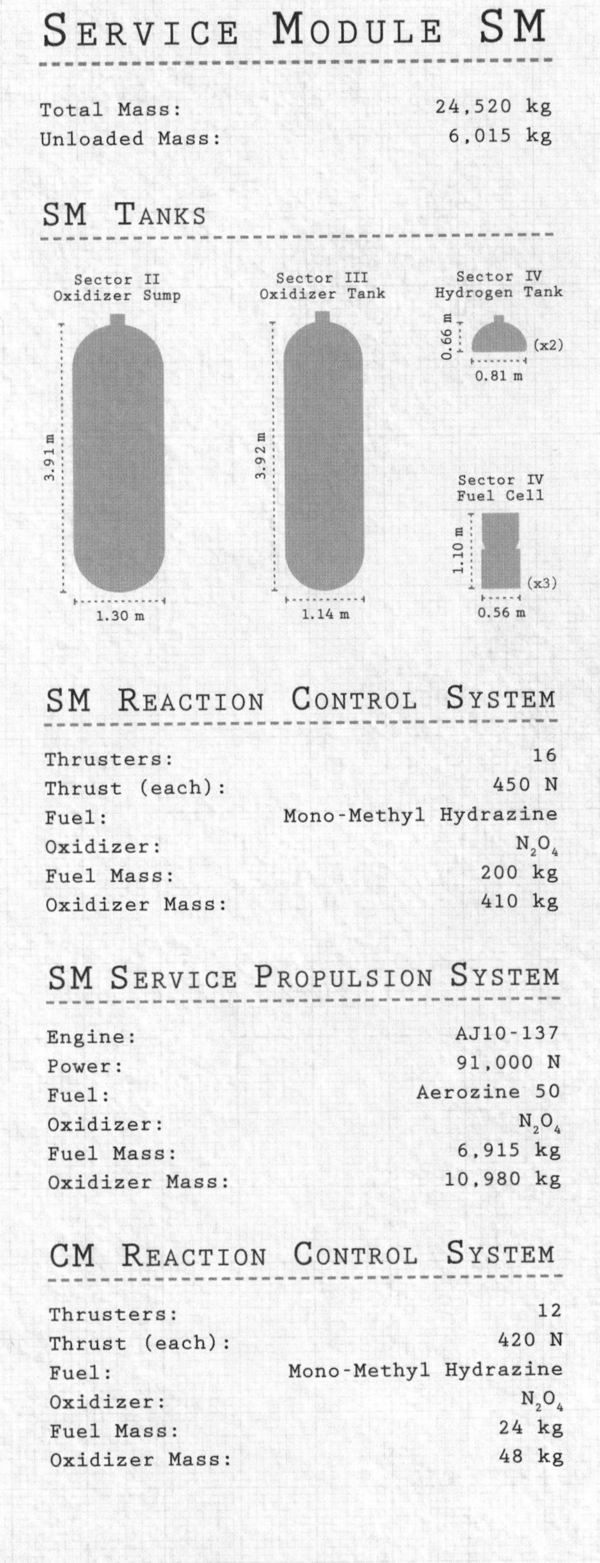

SM Reaction Control System

Thrusters:	16
Thrust (each):	450 N
Fuel:	Mono-Methyl Hydrazine
Oxidizer:	N_2O_4
Fuel Mass:	200 kg
Oxidizer Mass:	410 kg

SM Service Propulsion System

Engine:	AJ10-137
Power:	91,000 N
Fuel:	Aerozine 50
Oxidizer:	N_2O_4
Fuel Mass:	6,915 kg
Oxidizer Mass:	10,980 kg

CM Reaction Control System

Thrusters:	12
Thrust (each):	420 N
Fuel:	Mono-Methyl Hydrazine
Oxidizer:	N_2O_4
Fuel Mass:	24 kg
Oxidizer Mass:	48 kg

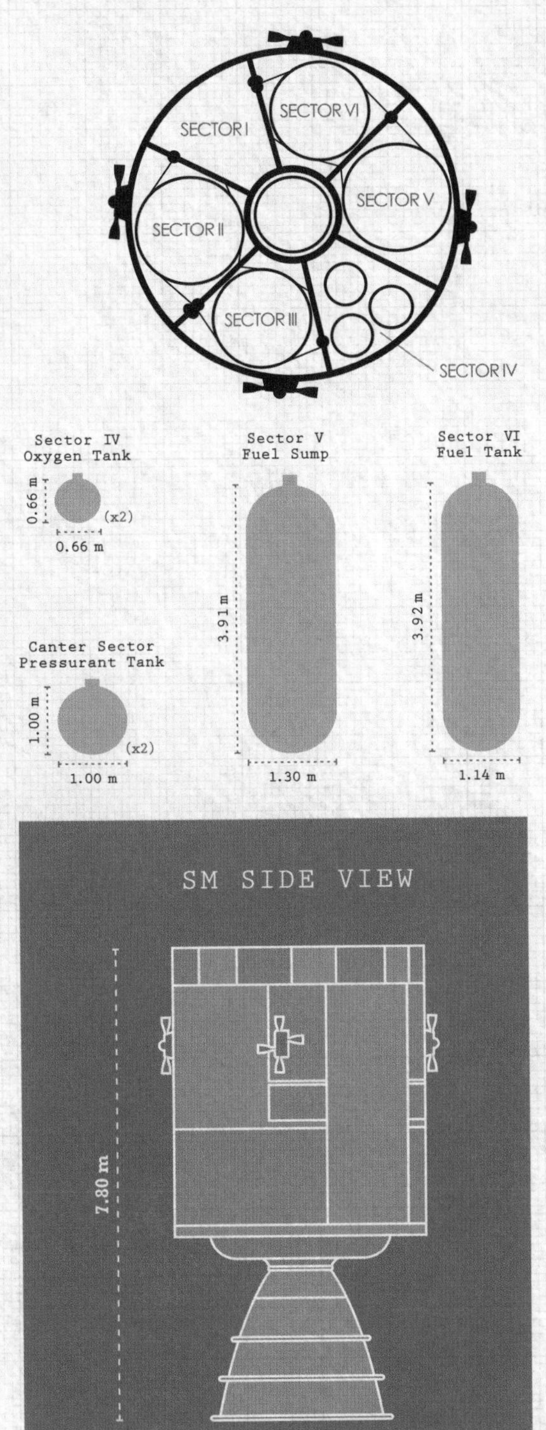

LUNAR MODULE LM

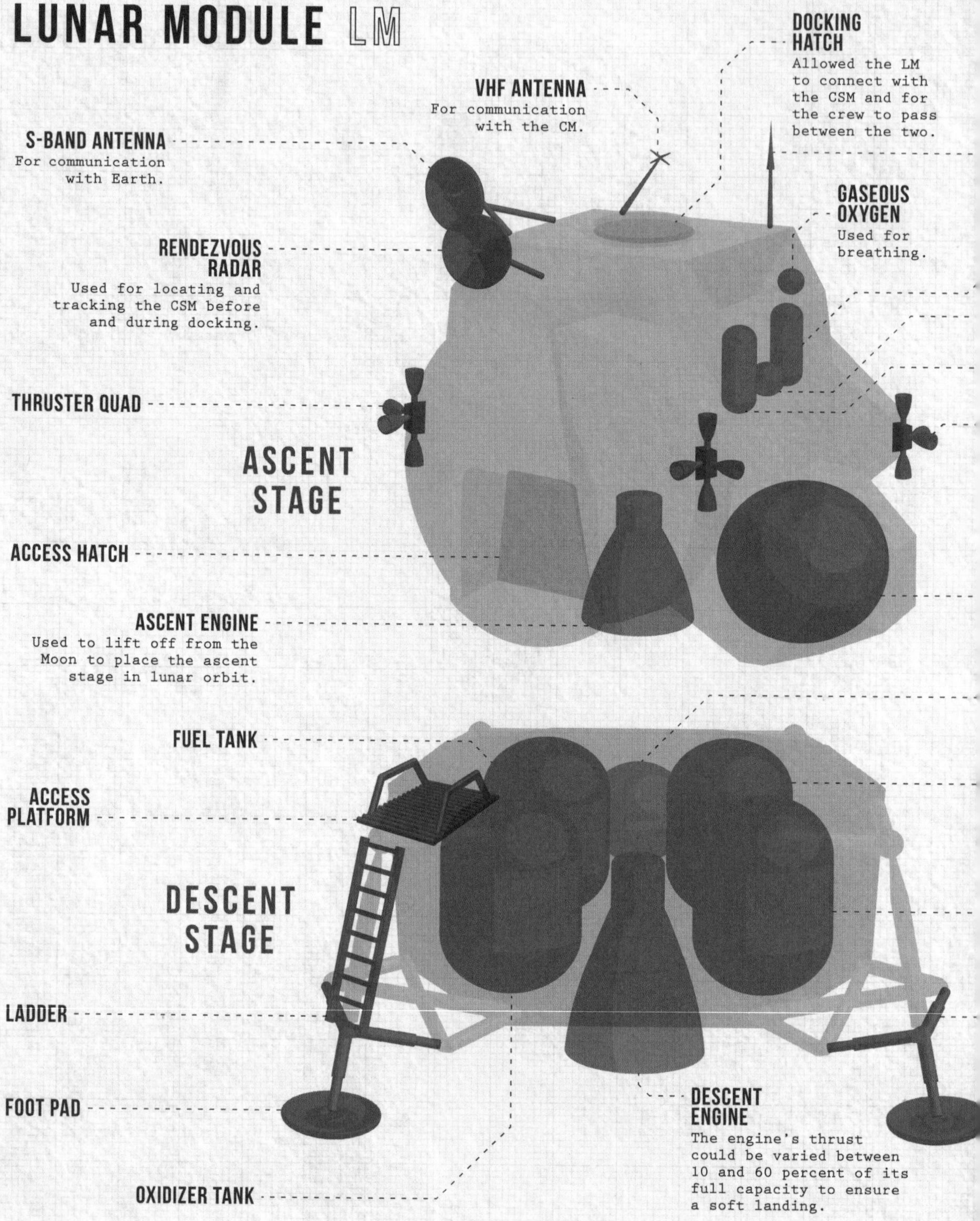

DOCKING HATCH
Allowed the LM to connect with the CSM and for the crew to pass between the two.

VHF ANTENNA
For communication with the CM.

S-BAND ANTENNA
For communication with Earth.

GASEOUS OXYGEN
Used for breathing.

RENDEZVOUS RADAR
Used for locating and tracking the CSM before and during docking.

THRUSTER QUAD

ASCENT STAGE

ACCESS HATCH

ASCENT ENGINE
Used to lift off from the Moon to place the ascent stage in lunar orbit.

FUEL TANK

ACCESS PLATFORM

DESCENT STAGE

LADDER

FOOT PAD

DESCENT ENGINE
The engine's thrust could be varied between 10 and 60 percent of its full capacity to ensure a soft landing.

OXIDIZER TANK

EVA ANTENNA
For communication between the LM and the astronauts outside.

RCS FUEL TANK

RCS PRESSURANT TANK

RCS OXIDIZER TANK

THRUSTER QUAD
Similar to the ones found on the SM, there were 16 thrusters situated on 4 thruster quads.

ASCENT FUEL TANK

PRESSURANT TANK

OXIDIZER TANK

FUEL TANK

SECONDARY SHOCK ABSORBER

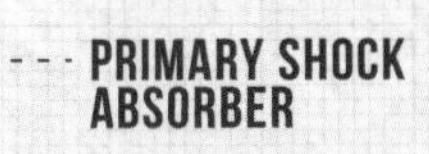

PRIMARY SHOCK ABSORBER

The LM was composed of two stages: descent and ascent. These stages remained together as a unit until it was time for the astronauts to leave the Moon's surface. The LM had four undercarriage legs that supported it during its stay on the Moon. These legs were folded in to save space while the craft was aboard the Saturn V, and would only be extended when the craft was in the Moon's orbit. The LM could only operate in the vacuum of space, as it was aerodynamically unsuitable to fly through the Earth's atmosphere.

The Descent Propulsion System (DPS) was housed in the descent stage. This slowed the craft down as it approached the Moon, and provided the astronauts with enough hover time to select a suitable landing site. The DPS consisted of the pressurant, fuel and oxidizer tanks, the engine, and all of their associated parts. On later missions, the Lunar Roving Vehicle (LRV) was also stowed in a compartment in the descent stage. Once the crew was ready to leave the Moon, this stage would become detached from the ascent stage and act as a launch platform. As a result the Apollo missions have left a total of six LM descent stages on the Moon's surface.

The ascent stage comprised the crew's cabin, which accommodated two personnel, as well as the LM's flight controls and instruments. Inside, crisscrossing hammocks were extended for the astronauts to sleep in during their rest breaks. It was pressurized like the CM and had four quads of RCS thrusters on its exterior that provided control. Should the SM thrusters fail, the LM thrusters were able to steer the spacecraft. The ascent stage also contained the Ascent Propulsion System (APS), which was required for lift-off from the descent stage at the end of a lunar visit. Once in orbit, after the crew transferred from the LM back into the CM, the LM's ascent stage would be jettisoned into solar orbit or to crash into the Moon.

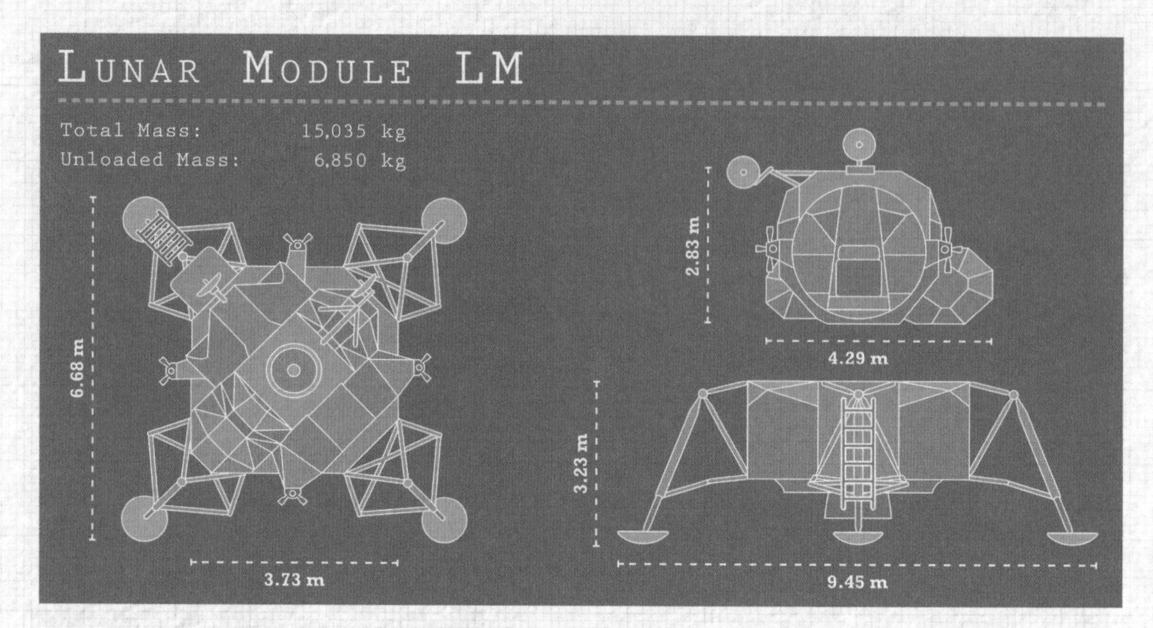

LM Ascent Stage

Crew Members:	2
Total Mass (inc. crew):	4,700 kg
Unloaded Mass:	2,150 kg
Habitable Volume:	4.50 m^3
Pressurized Volume:	6.70 m^3
Water Capacity:	38.6 kg
Batteries:	2
Total Battery Charge:	592 AH

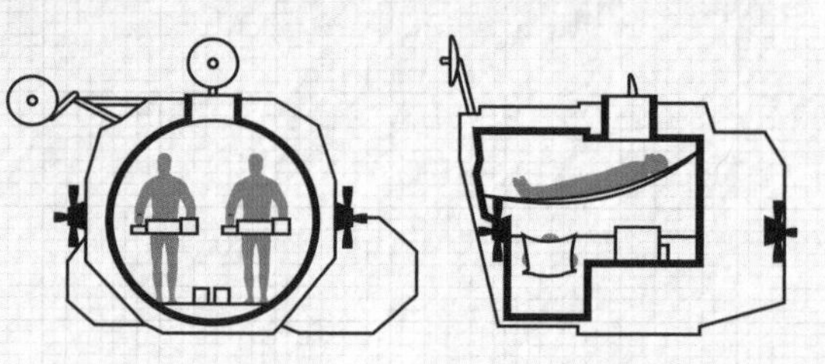

Ascent Propulsion System

Engine:	Bell LMAE
Thrust:	16,000 N
Fuel:	Aerozine 50
Oxidizer:	N_2O_4
Fuel Mass:	785 kg
Oxidizer Mass:	1,570 kg

Reaction Control System

Thrusters:	16
Thrust (each):	440 N
Fuel:	Mono-Methyl Hydrazine
Oxidizer:	N_2O_4
Fuel Mass:	97 kg
Oxidizer Mass:	190 kg

LM Ascent Stage Tanks

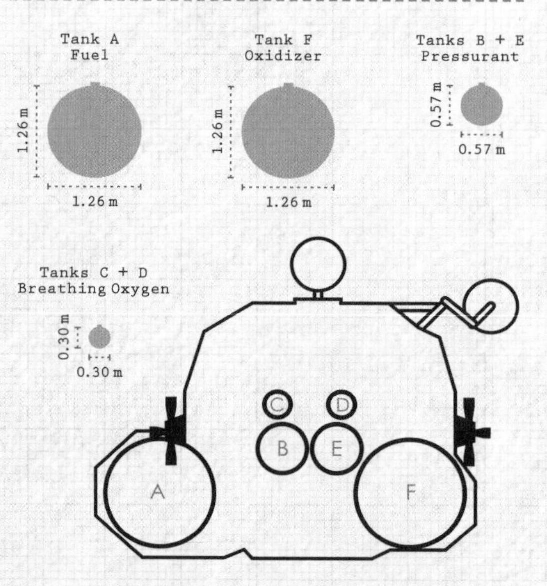

LM Descent Stage

Total Mass:	10,335 kg
Unloaded Mass:	2,150 kg
Batteries:	2
Total Battery Charge:	592 AH
Water Capacity:	150 kg
Undercarriage Legs:	4
Footpad Diameter:	0.91 m

Descent Propulsion System

Engine:	TRW LMDE
Thrust:	45,040 N
Fuel:	Aerozine 50
Oxidizer:	N_2O_4
Fuel Mass:	2,735 kg
Oxidizer Mass:	5,300 kg

LM Descent Stage Tanks

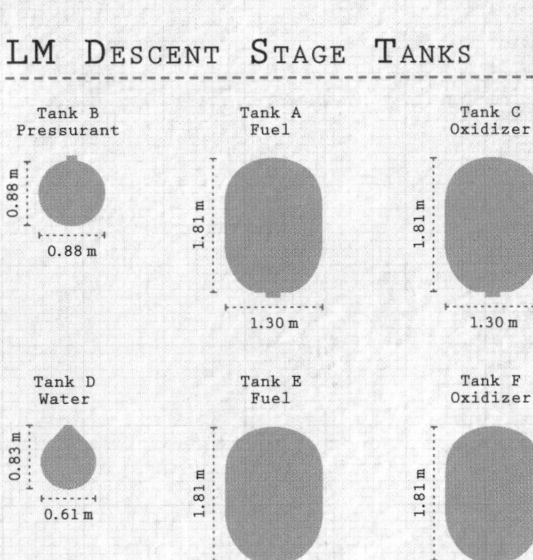

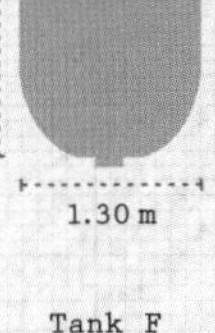

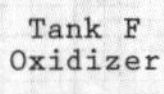

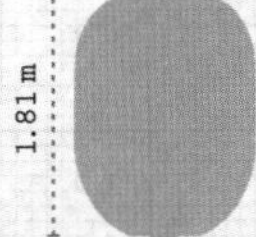

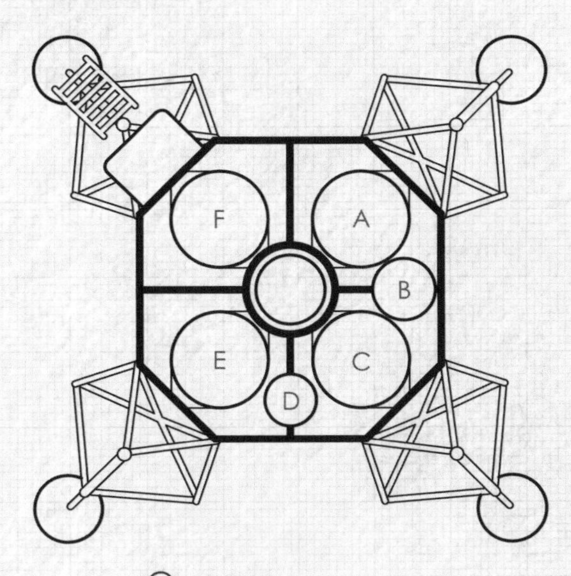

LANDING GEAR IN STOWED POSITION

6.06 m

LUNAR ROVING VEHICLE LRV

The Lunar Roving Vehicle (LRV), popularly known as the "Moon buggy," was first deployed in 1971 on Apollo 15. Four LRVs were built in total, with two employed on Apollos 16 and 17, and the remaining vehicle used for spare parts. The LRV's purpose was to allow the astronauts to explore a much greater area of the Moon than they could on foot. On earlier lunar landing missions, astronauts were limited to short walking distances around the LM due to their cumbersome spacesuits and equipment.

The LRV was a four-wheel drive, electrically powered vehicle designed to have a top speed of 13 km/h (though it achieved 14 km/h). Capable of carrying a payload of more than double its own mass, it could comfortably transport two astronauts, their scientific equipment, and rock samples over the lunar surface. Unlike a conventional car, it was controlled by a single T-shaped hand controller rather than a steering wheel and pedals. The controller operated the four drive motors and the two steering motors. Front- and rear-wheel steering helped it to achieve a tight turning circle. The LRV could be put into reverse by flipping the switch on the stem of the controller before pulling the stick backward. The stick was pulled all the way back to activate the parking brake. A color TV camera was mounted to the front of the LRV, which allowed for much better television coverage than on previous missions.

The LRVs proved to be dependable, with only a couple of very minor exceptions. Designed, built, and tested in a relatively short period of seventeen months, the LRV is credited with making the major scientific discoveries of Apollos 15 to 17 possible. The three LRVs that were taken to the Moon remain there to this day; just like the descent stage of the LM, the LRV would be left on the Moon's surface when the astronauts returned to Earth.

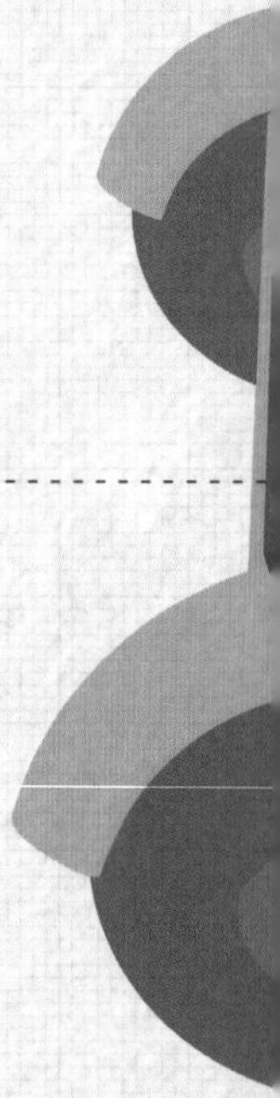

HAND CONTROLLER

This stick situated between the two seats controlled the four drive motors, two steering motors, and brakes.

BUDDY SECONDARY LIFE SUPPORT SYSTEM

A set of hoses and connectors that allowed the astronauts to share cooling water between their portable life support systems, should one of them fail.

REAR PAYLOAD PALLET

Tools and scientific equipment were stored here, as well as the rock samples.

LOW-GAIN ANTENNA
Provided voice communications between Houston and the crew while driving.

HIGH-GAIN ANTENNA
This antenna was used for TV transmissions.

INSTRUMENT PANEL
This provided information on speed, heading, temperature levels, and pitch.

TELEVISION CAMERA
This camera could be remotely operated by Mission Control.

TELEVISION CONTROL UNIT

LUNAR COMMUNICATION RELAY UNIT

16MM CAMERA
Used for data acquisition.

WHEELS
The treads were made from interwoven, zinc-coated steel strands. Strips of titanium plate covered 50 percent of the contact area to aid with traction.

UNDER-SEAT STORAGE

Lunar Roving Vehicle LRV

STOWED TOP VIEW

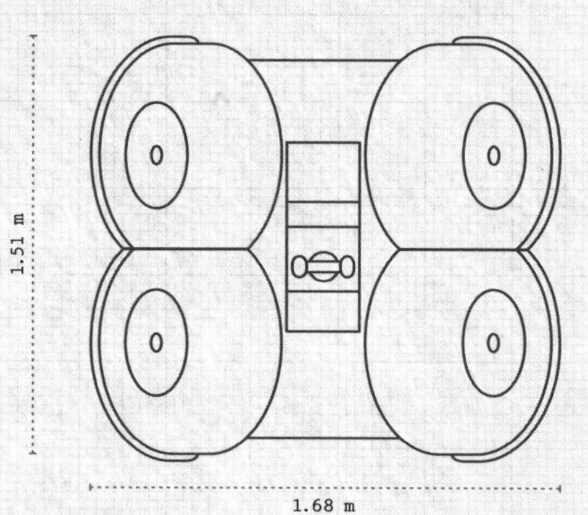

STOWED END VIEW

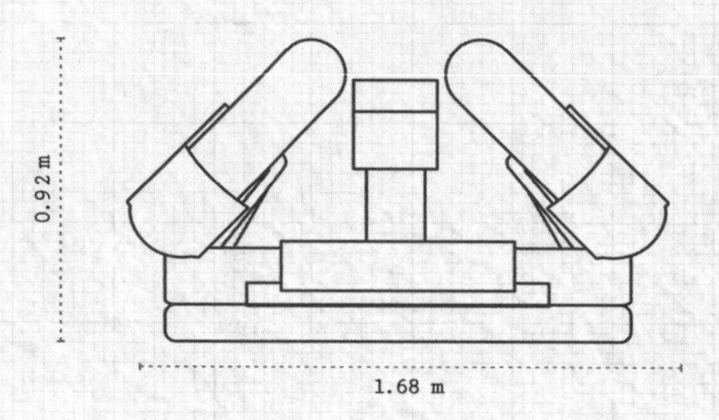

Loaded Mass:	726 kg
Unloaded Mass:	210 kg
Maximum Speed:	14 km/h
Wheel Base:	2.3 m
Ground Clearance:	0.35 m
Turning Radius:	3.05 m
Batteries:	2
Total Battery Charge:	242 AH
Voltage:	36 V
Range:	92 km

LRV Wheels

Wheels:	4
Power (each):	190 W
Mass (each):	5.4 kg
Diameter:	0.81 m
Width:	0.23 m

Lunar Roving Vehicle Stowage Folding Sequence

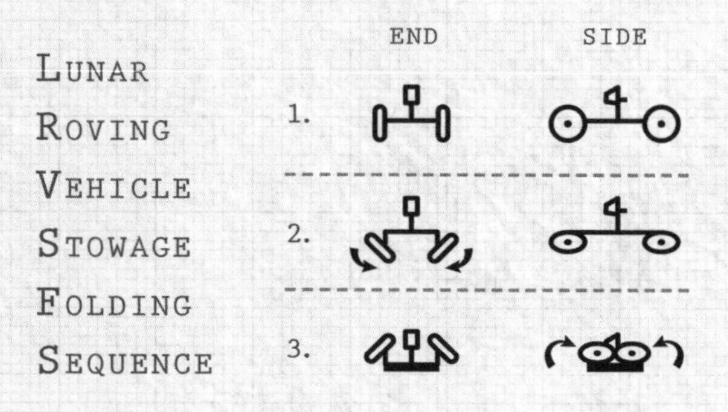

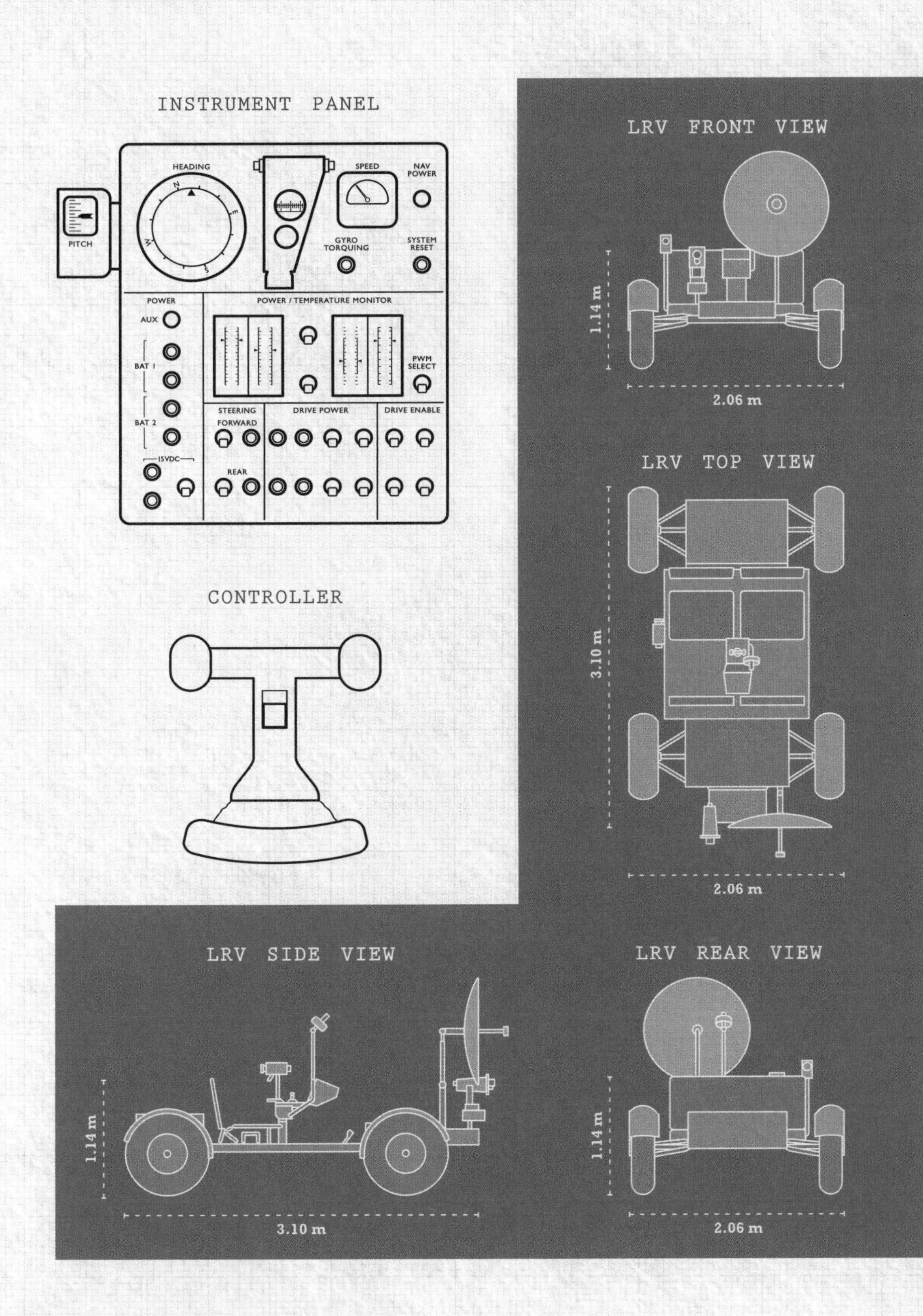
INSTRUMENT PANEL
HEADING
SPEED
NAV POWER
PITCH
GYRO TORQUING
SYSTEM RESET
POWER
POWER / TEMPERATURE MONITOR
AUX
BAT 1
BAT 2
PWM SELECT
STEERING
FORWARD
DRIVE POWER
DRIVE ENABLE
15VDC
REAR
CONTROLLER
LRV FRONT VIEW
1.14 m
2.06 m
LRV TOP VIEW
3.10 m
2.06 m
LRV SIDE VIEW
1.14 m
3.10 m
LRV REAR VIEW
1.14 m
2.06 m

SATURN V ROCKET SAT V

The Saturn V would hold the record as the largest and most powerful operational rocket for nearly fifty years. In total thirteen Saturn Vs were launched between 1966 and 1973, carrying out Apollo missions as well as placing the Skylab space station in orbit. Each Saturn V assembly cost roughly $185 million (more than $1 billion today), and every part of the massive rocket was discarded during the mission.

FIRST STAGE S-IC

The S-IC was the first stage of the Saturn V. Its five huge F1 engines fired the rocket from the launchpad until it was 68 km above the Earth travelling at 9,200 km/h. When all the fuel had been used, the engines would shut down and explosive bolts would fire, separating the first stage away from the interstage adapter. The S-IC would then fall into the ocean, at least 500 km away from the launch site.

FUEL TANK

Rocket Propellant-1 (RP1) was the fuel used in the first stage. It is a highly refined form of kerosene, similar to jet fuel. It provided less of an explosive hazard than liquid hydrogen and produced more power by volume.

INTERSTAGE ADAPTER

Contained eight small rocket motors that separated the adapter from the second stage of the rocket.

F1 ROCKET ENGINES

The F1 was the most powerful single-nozzle liquid-fuelled rocket engine ever built. The central engine was rigidly mounted, while the outer four were gimballed to keep the rocket straight when it was tearing through the atmosphere. Each one burned 1,565 liters of oxidizer and 976 liters of fuel every second, and was able to withstand the extreme range of temperatures it was subjected to.

OXIDIZER TANK

Stored the liquid oxygen used as oxidizer in many rocket engines.

SECOND STAGE S-II

The second stage accelerated the spacecraft through the upper atmosphere, taking the vehicle and payload to an altitude of 175 km and a speed of 24,600 km/h—close to orbital velocity. It had a burn time of 6 minutes, after which it was fired away from the third stage, impacting Earth more than 4,000 km from its point of origin.

THIRD STAGE S-IVB

This stage fired twice. After the second stage separated, the third stage fired for two minutes and thirty-five seconds to place the craft into a safe orbit of Earth at 28,000 km/h. It would then fire again to pull out of Earth orbit toward the Moon at a speed of more than 39,000 km/h. After the S-IVB was spent, it was either sent into solar orbit or to crash into the Moon's surface.

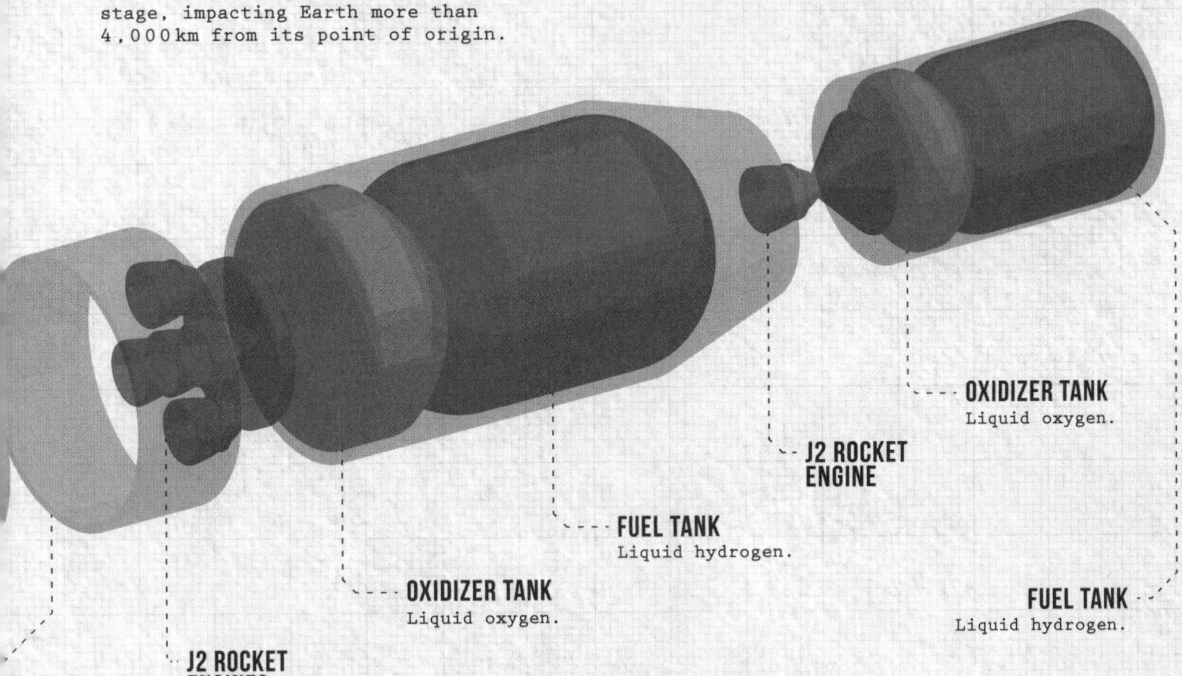

OXIDIZER TANK
Liquid oxygen.

J2 ROCKET ENGINE

FUEL TANK
Liquid hydrogen.

OXIDIZER TANK
Liquid oxygen.

FUEL TANK
Liquid hydrogen.

J2 ROCKET ENGINES
Each of the five engines generated 1,033,000 N thrust in the vacuum of space, and just less than half of that at sea level. They underwent several upgrades during their operational history.

The rocket was designed by Wernher von Braun, who at the time was heading a team of scientists in the rocket design division of the Army. Overall, the Saturn V performed reliably, although on two occasions, Apollos 6 and 13, it experienced engine loss during launch. It was, however, able to recover the mission by having the remaining engines burn for longer to compensate. Throughout its seven-year service history, the Saturn V never lost a payload.

Saturn V Rocket

S-IVB stage 3

Loaded Mass:	123,000 kg
Unloaded Mass:	13,500 kg
Engine Type:	J2
No. of Engines:	1
Thrust:	1,001 kN
Oxidizer:	Liquid O_2
Fuel:	Liquid H_2
Oxidizer Mass:	90,500 kg
Fuel Mass:	19,000 kg
Burn Time:	165 + 335 seconds (2 burns)

S-II stage 2

Loaded Mass:	480,900 kg
Unloaded Mass:	53,900 kg
Engine Type:	J2
No. of Engines:	5
Thrust:	4,400 kN
Oxidizer:	Liquid O_2
Fuel:	Liquid H_2
Oxidizer Mass:	358,000 kg
Fuel Mass:	69,000 kg
Burn Time:	367 seconds

S-IC stage 1

Loaded Mass:	2,280,000 kg
Unloaded Mass:	190,000 kg
Engine Type:	F1
No. of Engines:	5
Thrust:	33,850 kN
Oxidizer:	Liquid O_2
Fuel:	RP-1
Oxidizer Mass:	1,440,000 kg
Fuel Mass:	650,000 kg
Burn Time:	150 seconds

SAT-V complete

Loaded Mass:	2,883,900 kg
Unloaded Mass:	257,400 kg

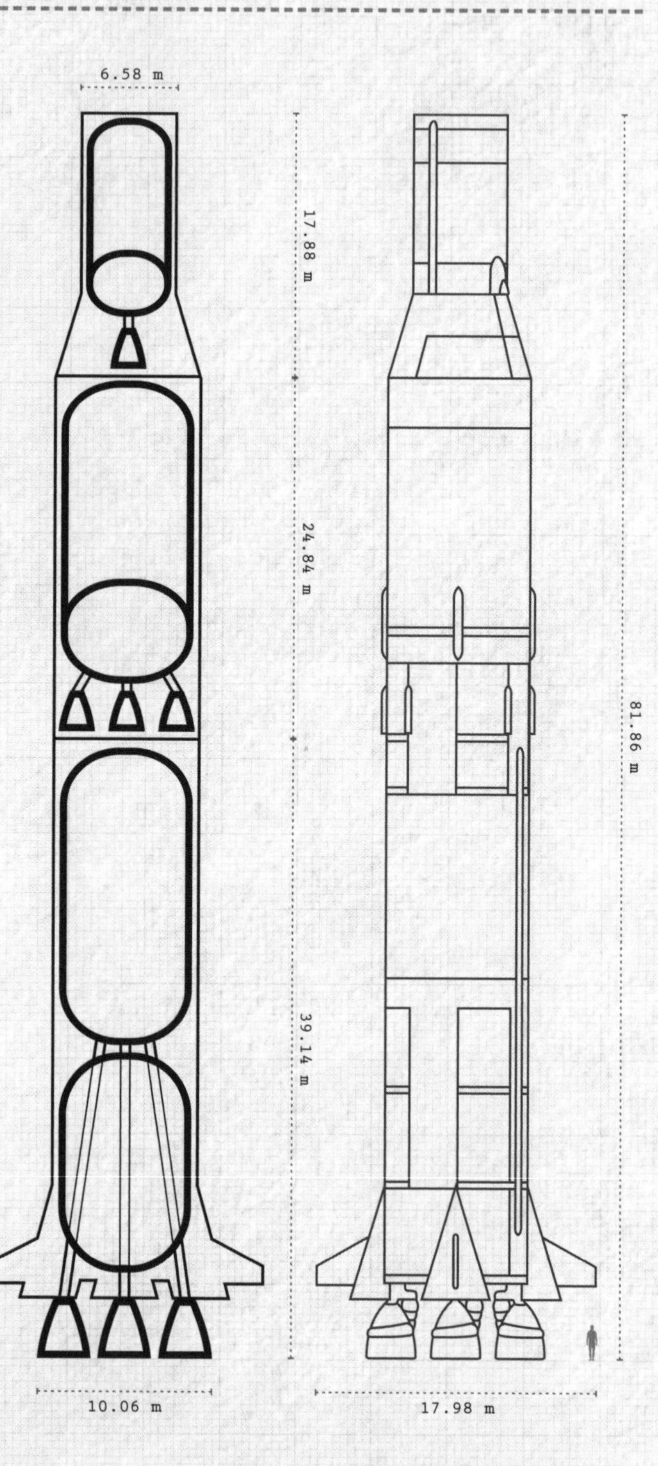

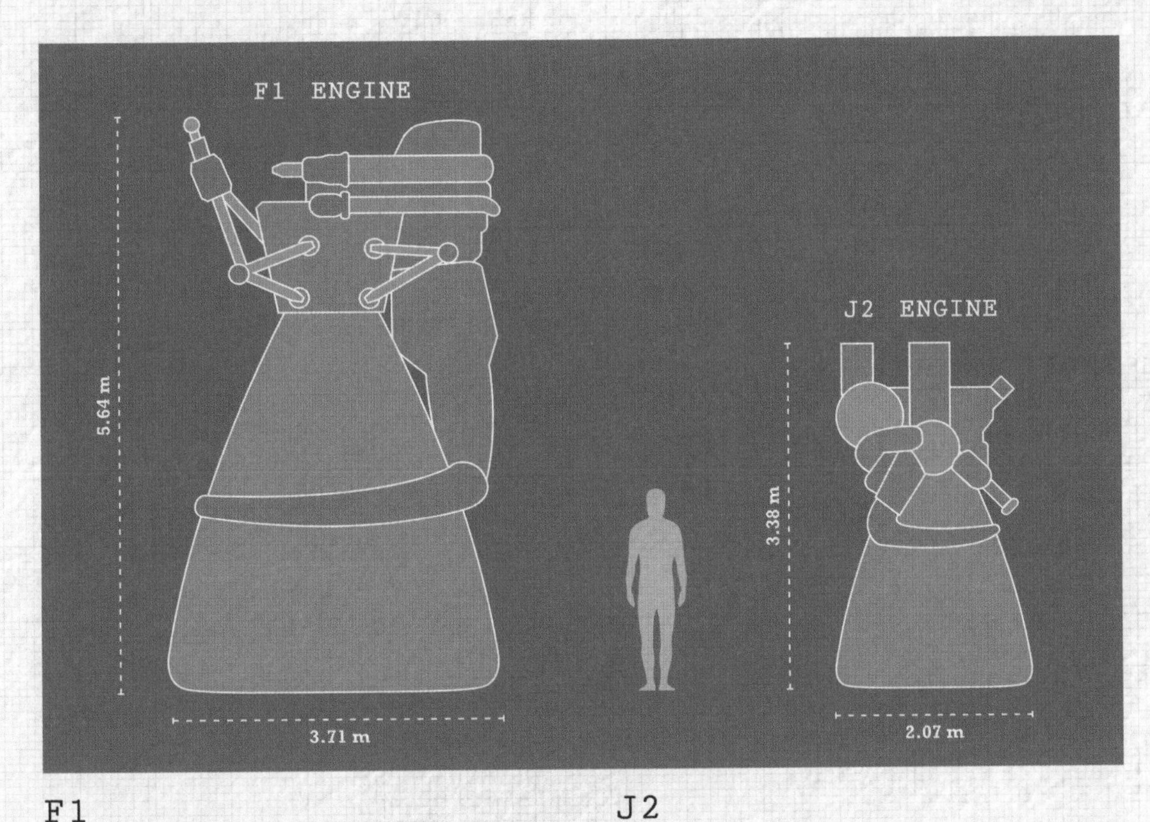

F1

Mass (dry):	8,400 kg
Thrust:	6,770 kN
Chamber Pressure:	6,650 kPa
Combustion Temperature:	3,300 ºC
Oxidizer:	Liquid O_2
Fuel:	RP-1
Oxidizer Flow Rate:	1,790 kg/s
Fuel Flow Rate:	788 kg/s

J2

Mass (dry):	1,438 kg
Thrust:	1,033 kN
Chamber Pressure:	5,260 kPa
Combustion Temperature:	3,180 ºC
Oxidizer:	Liquid O_2
Fuel:	Liquid H_2
Oxidizer Flow Rate:	204 kg/s
Fuel Flow Rate:	37 kg/s

Saturn V Engine Configuration

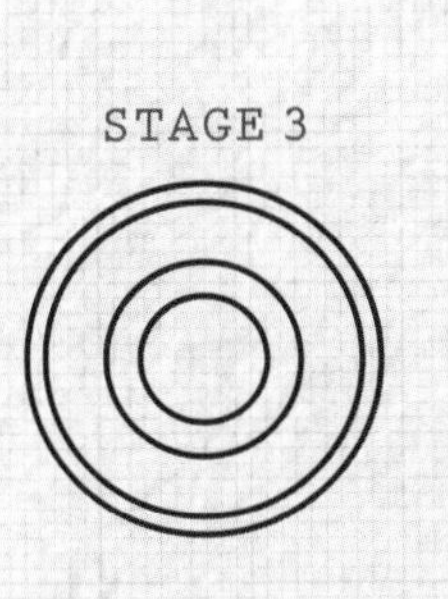

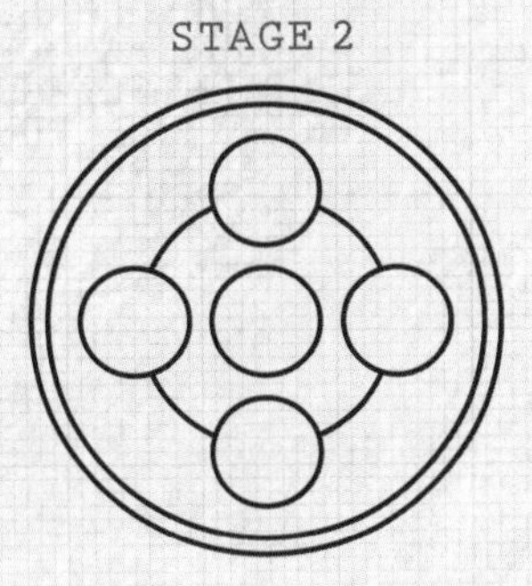

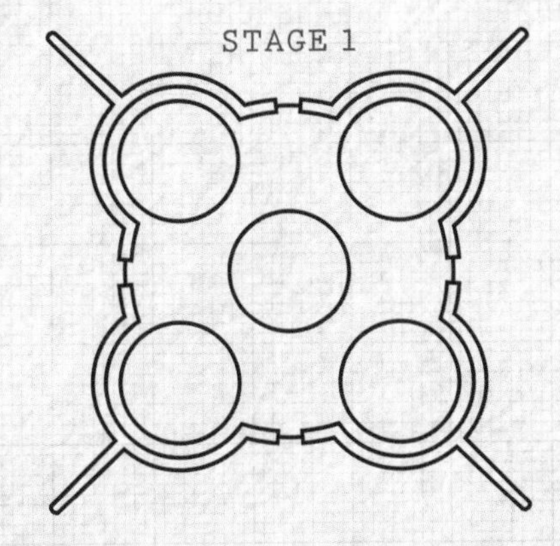

A7L SPACESUIT

VHF ANTENNA
Connected to the radio in the PLSS.

VISOR
This outer visor was worn while on the Moon's surface. It was coated with a thin layer of gold that reflected harmful radiation and helped to maintain a comfortable temperature within the helmet.

PRESSURE HELMET
Made from high-strength polycarbonate, the famous "fishbowl helmet" granted astronauts an unrestricted view. From Apollo 13 onward, a drinking straw that led to a water pouch was fitted inside. This enabled the astronauts to perform extended EVAs.

PLSS CONTROL UNIT
Allowed for monitoring and control of the fluids and electricity distributed by the PLSS.

COMMUNICATION UMBILICAL

OXYGEN PURGE SYSTEM

OXYGEN PURGE VALVE

BIOMEDICAL ACCESS FLAP
Underneath this cover there was an injection patch, used to administer shots as directed by NASA doctors. It was designed to be self-sealing, although luckily it was never needed. The flap also covered the urine transfer connector.

UTILITY POCKET

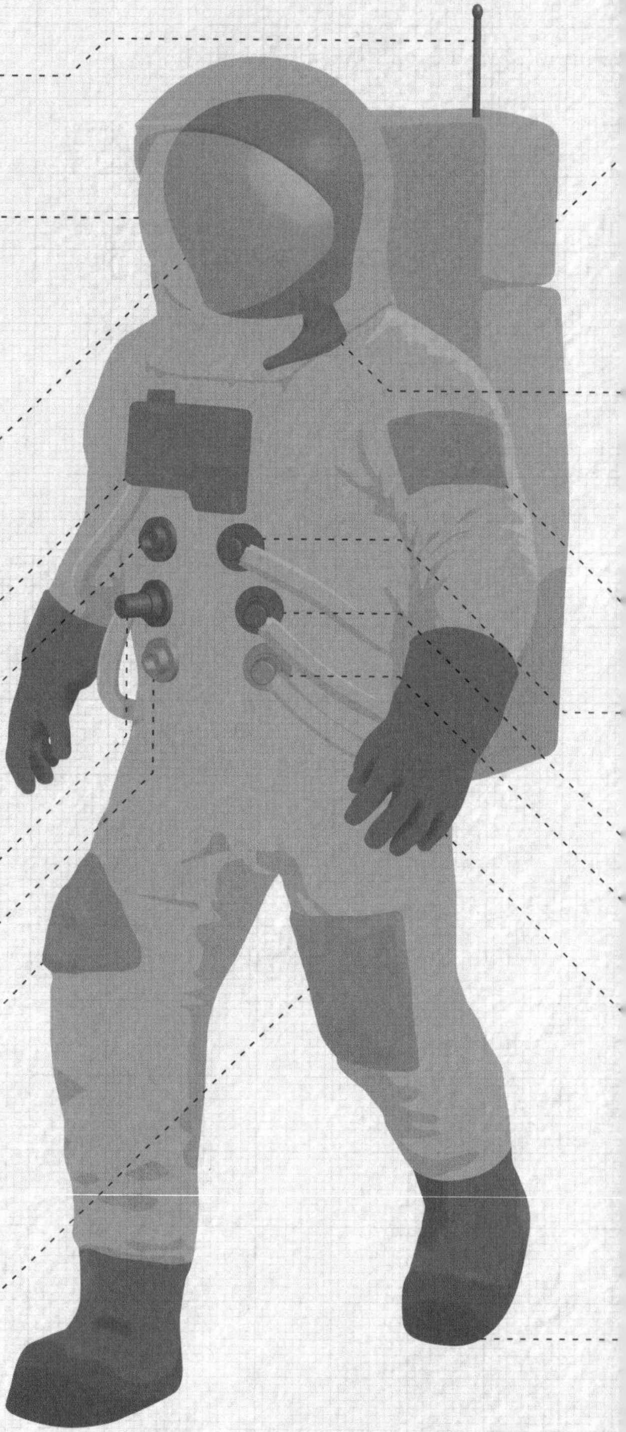

PORTABLE LIFE SUPPORT SYSTEM
Worn like a backpack, the PLSS contained oxygen tanks, water-cooling equipment, a two-way radio, and a battery for electrical power. These services were all supplied to the spacesuit through umbilical tubes. The top section of the backpack contained the Oxygen Purge System, which could provide breathing oxygen should the primary system fail.

DRINK BAG
This was worn underneath the neck ring of the pressure suit. A straw led from the bag to the astronaut's mouth, with a valve at the end to prevent leakage inside the helmet.

SUNGLASSES POCKET

WATER CONNECTOR
Cooling water was transferred both into and out of the spacesuit through this one umbilical.

OXYGEN INLET

OXYGEN OUTLET

GLOVES
The multilayered inner gloves, attached to the pressure suit, were taken from molds of each astronaut's hands. This ensured they had the sensitivity to operate controls and tools. The outer gloves were much more durable and provided thermal insulation.

LUNAR OVERBOOTS
The outer layer of the boot was durable, made from metal-woven fabric with a ribbed rubber sole.

The spacesuits used on the Apollo missions were based on the suits the astronauts wore for Project Gemini; however, they were significantly updated. The A7L was worn on all the Apollo flights, with modified versions used on later missions, including on Apollo–Soyuz and Skylab.

The suit itself consisted of five main layers. Next to the astronauts' skin was the liquid-cooled garment, which was a type of bodysuit with narrow tubing sewn in. Then there was a layer of nylon that provided some comfort, followed by a pressure bladder that made it easier for the astronauts to bend their joints in the pressurized suits. On the outside of the bladder there was another layer of nylon to help keep it in place. A special zip ran down the back, from the shoulders to the waist. This not only allowed for access, but also retained pressure inside the suit. The final layer was a cover, worn over the pressure suit. It was designed to provide heat insulation, to protect against wear and tear, and to protect against micrometeorites. The helmet and gloves were attached to the suit by pressure-sealing metal rings, whereas the boots were simply protective overshoes that slipped on over the spacesuit's integral pressure boot. While on the Moon, an outer visor, whose prime function was to shield the astronaut's eyes from harmful UV radiation, was worn over the helmet.

The backpack was known as the Portable Life Support System (PLSS), and it allowed the astronauts to survive away from their spacecraft. It supplied them with breathable oxygen, as well as cooling and circulating oxygen through the pressure bladders in their suit, and cooling and circulating water through the liquid-cooled garment. It also provided them with two-way voice communication. The A7L was custom-made for every Apollo astronaut, each of whom received three: one for flight, one for training, and one for backup.

A7L Spacesuit

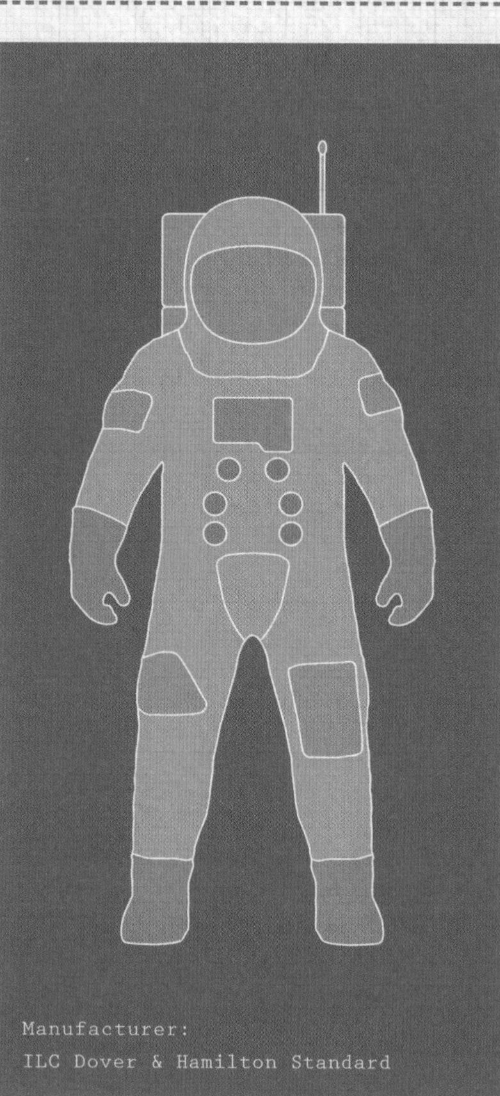

Helmet

Material: Polycarbonate

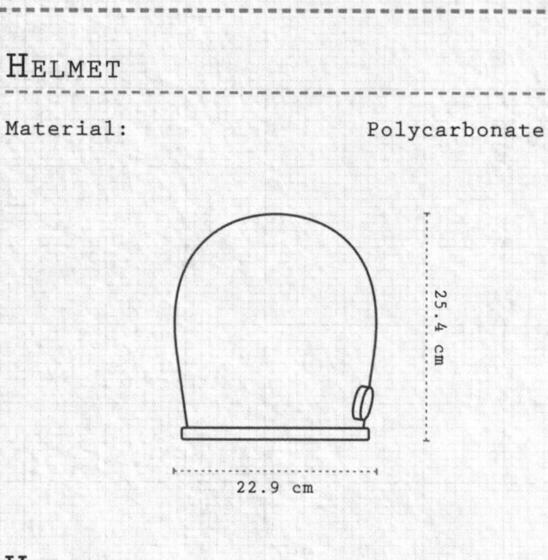

Visor

Exterior: Gold-laminated polycarbonate
Interior: UV Plexiglass

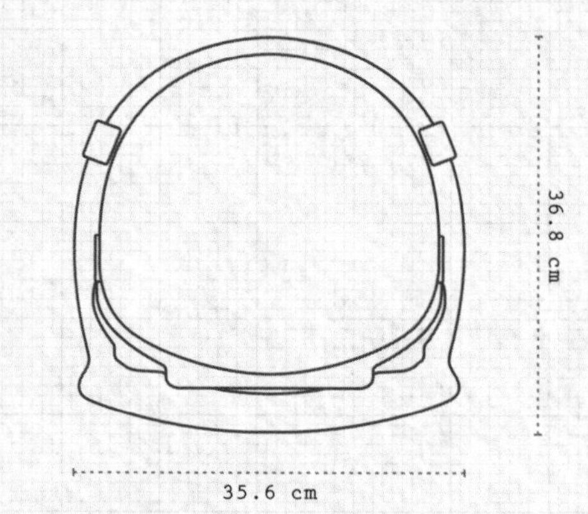

Drink Bag

Volume: 0.6 L

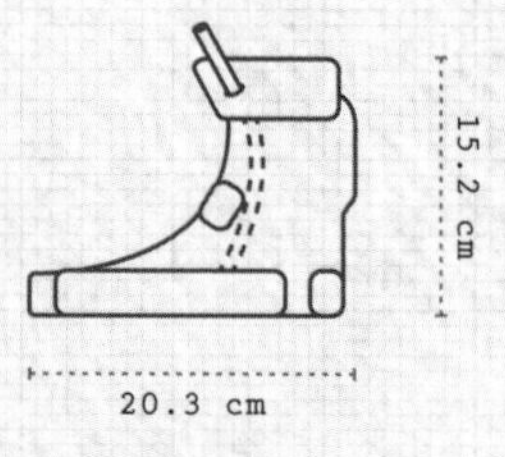

Suit

Operating Pressure:	25.5 kPa
Pressurized Volume:	0.12-0.15 m^3
Free Pressurized Volume (when worn):	0.06 m^3
Primary Life Support:	6 hrs
Back-up Life Support:	30 mins
IVA Suit Mass:	28.1 kg
EVA Suit Mass:	34.5 kg

PORTABLE LIFE SUPPORT SYSTEM PLSS

PLSS

PLSS Mass: 28.4 kg
Oxygen Tank Pressure: 7,000 kPa
Battery: 279 Watt-hour
Cooling Water: 3.9 L

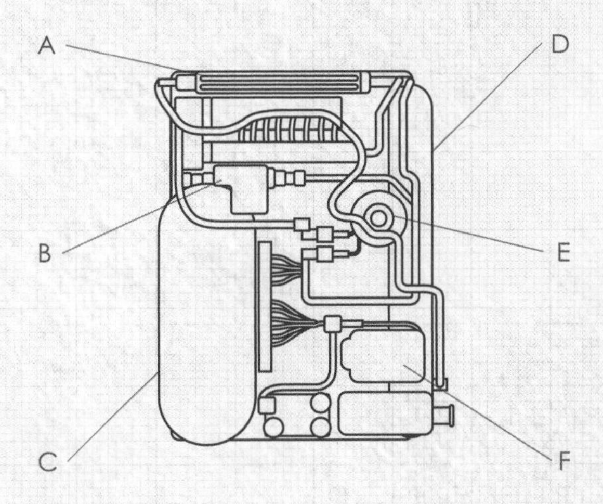

A:........................Sublimator
B:.............Ventilation Flow Sensor
C:................Primary Oxygen Bottle
D:...........Auxiliary Water Reservoir
E:..................Pressure Transducer
F:............................Battery

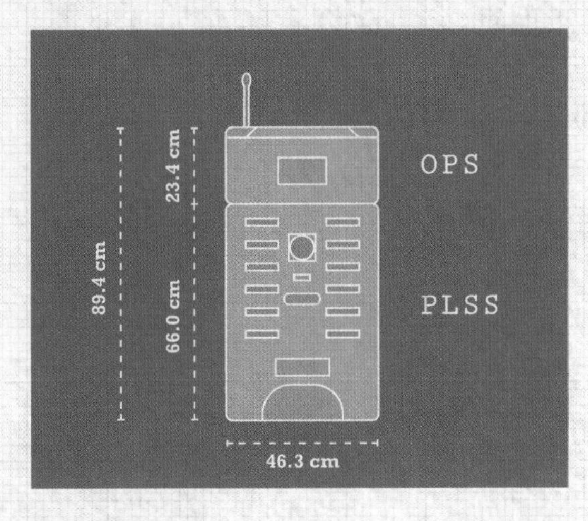

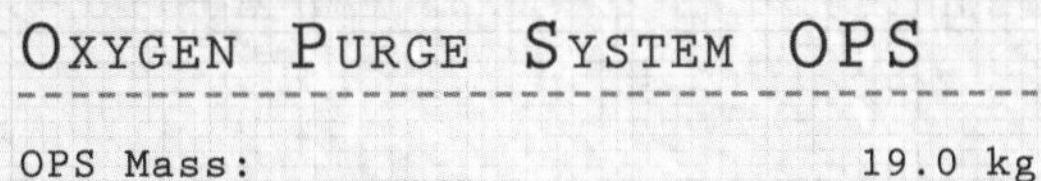

OXYGEN PURGE SYSTEM OPS

OPS Mass: 19.0 kg

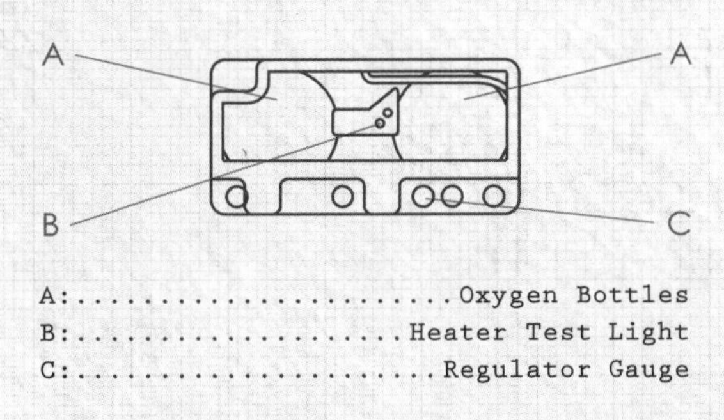

A:......................Oxygen Bottles
B:...................Heater Test Light
C:....................Regulator Gauge

PLSS CONTROL UNIT

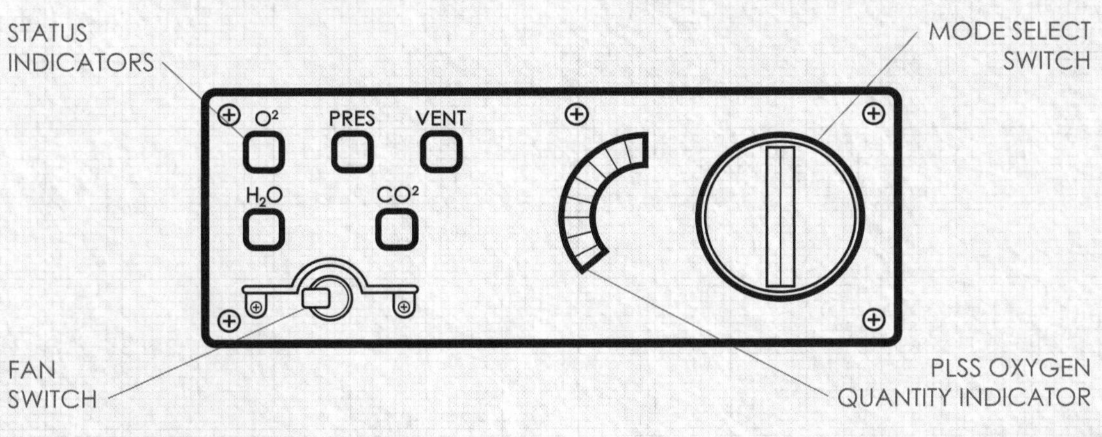

CRAWLER-TRANSPORTER

Because the Saturn rocket was so enormous, an equally large machine, the Crawler-Transporter, had to be created to transport it to the launch pad. Upon construction the Crawler-Transporter became the largest self-powered land vehicle in the world. Its top deck was square and flat, with an area greater than a soccer field. The eight tracks it drove on were each bigger than a bus, and it weighed more than 2,700 tonnes in total. It was controlled from one of the two control cabs at either end of the vehicle, and each side of its deck could be raised independently so that the rocket it was carrying would remain vertical when travelling over inclines. Two Crawler-Transporters were built and both are still operational.

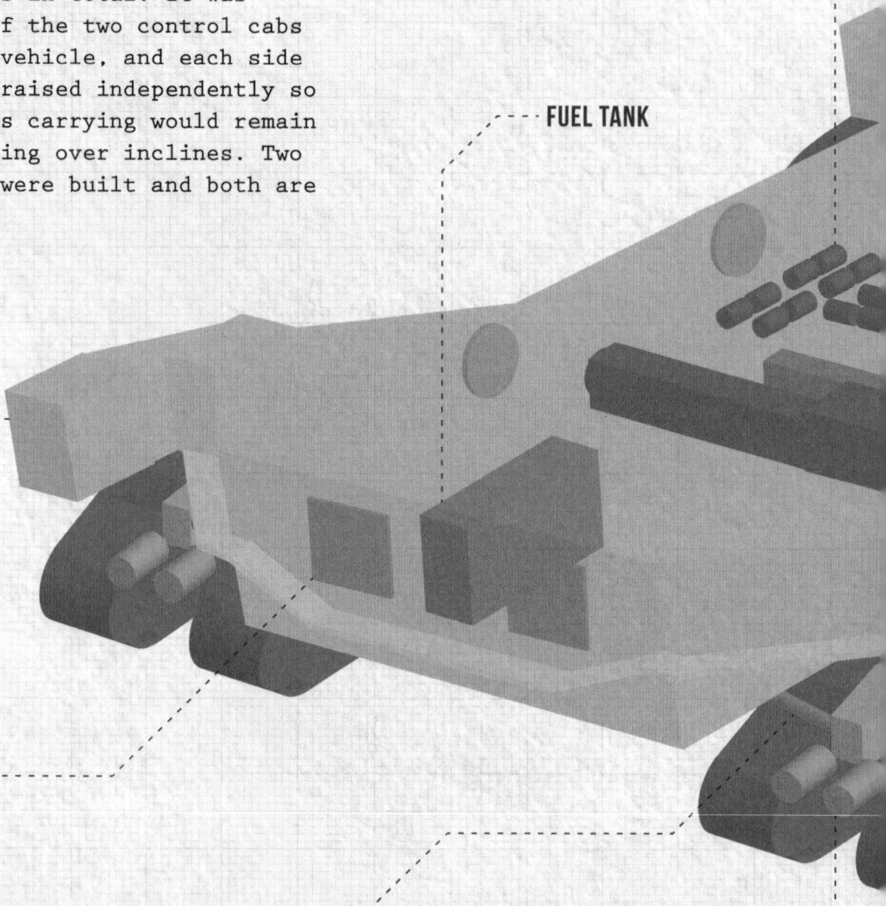

HYDRAULIC PUMPS
There were 4 pumps integral to the steering system and 8 pumps dedicated to jacking, equalizing, and leveling.

FUEL TANK

CONTROL CAB
The vehicle was operated from either end, in one of the two control cabs.

WATER COOLING RADIATORS

STEERING CYLINDER
These hydraulic arms connected the chassis to each end of the trucks, providing the force needed to turn the vehicle. There were 8 in total.

TRACTION MOTORS
It took 2 traction motors to turn each of the 8 tracks, and so 16 were fitted to the vehicle.

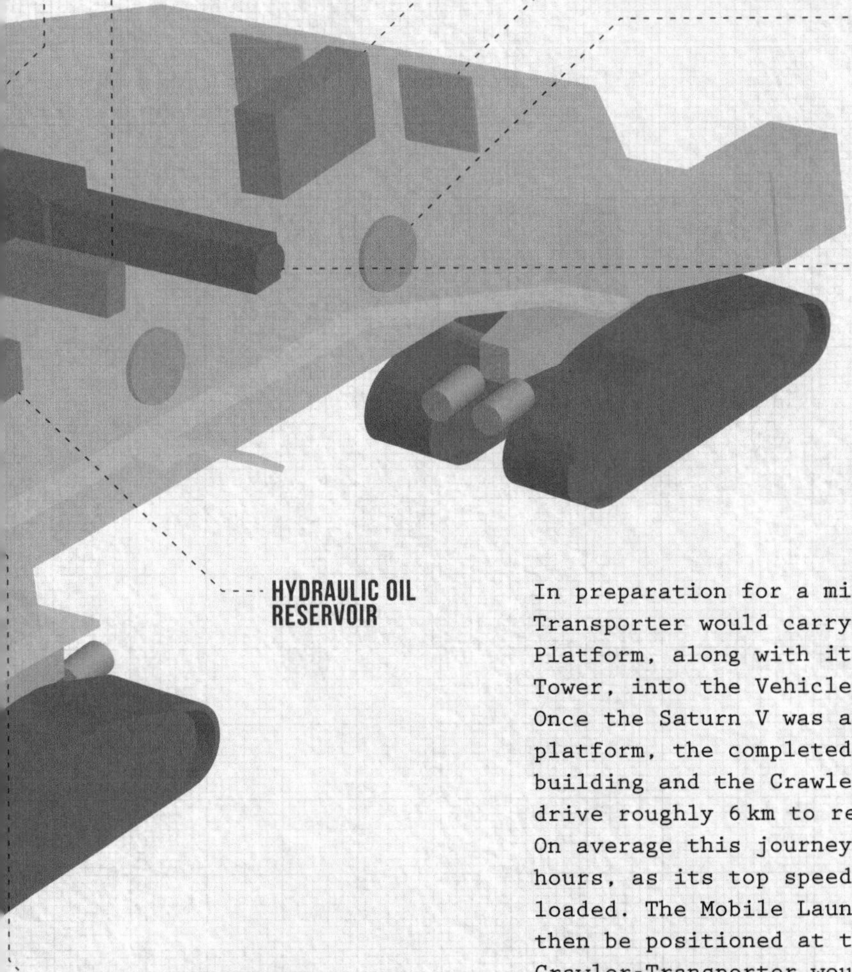

VENTILATING FANS

FUEL TANK

750 KW AC ENGINE GENERATOR SET
Two of these diesel engine generator sets were used to power the steering, jacking, lighting, and ventilation.

WATER RADIATORS

VENTILATING FANS

2,000 KW DC ENGINE GENERATOR SET
These were the largest engines on the vehicle, with one at either end. They ran on diesel and each powered two 1,000 kW generators, which in turn powered the traction motors.

HYDRAULIC OIL RESERVOIR

150KW AC GENERATOR
These generators were available to provide power to the Mobile Launcher Platform.

In preparation for a mission, the Crawler-Transporter would carry the Mobile Launcher Platform, along with its Launch Umbilical Tower, into the Vehicle Assembly Building. Once the Saturn V was assembled on the platform, the completed stack left the building and the Crawler-Transporter would drive roughly 6 km to reach the launch pad. On average this journey took around five hours, as its top speed was just 1.6 km/h when loaded. The Mobile Launcher Platform would then be positioned at the launch pad, and the Crawler-Transporter would move back to a safe location. After launch the transporter would return the platform to the Vehicle Assembly Building.

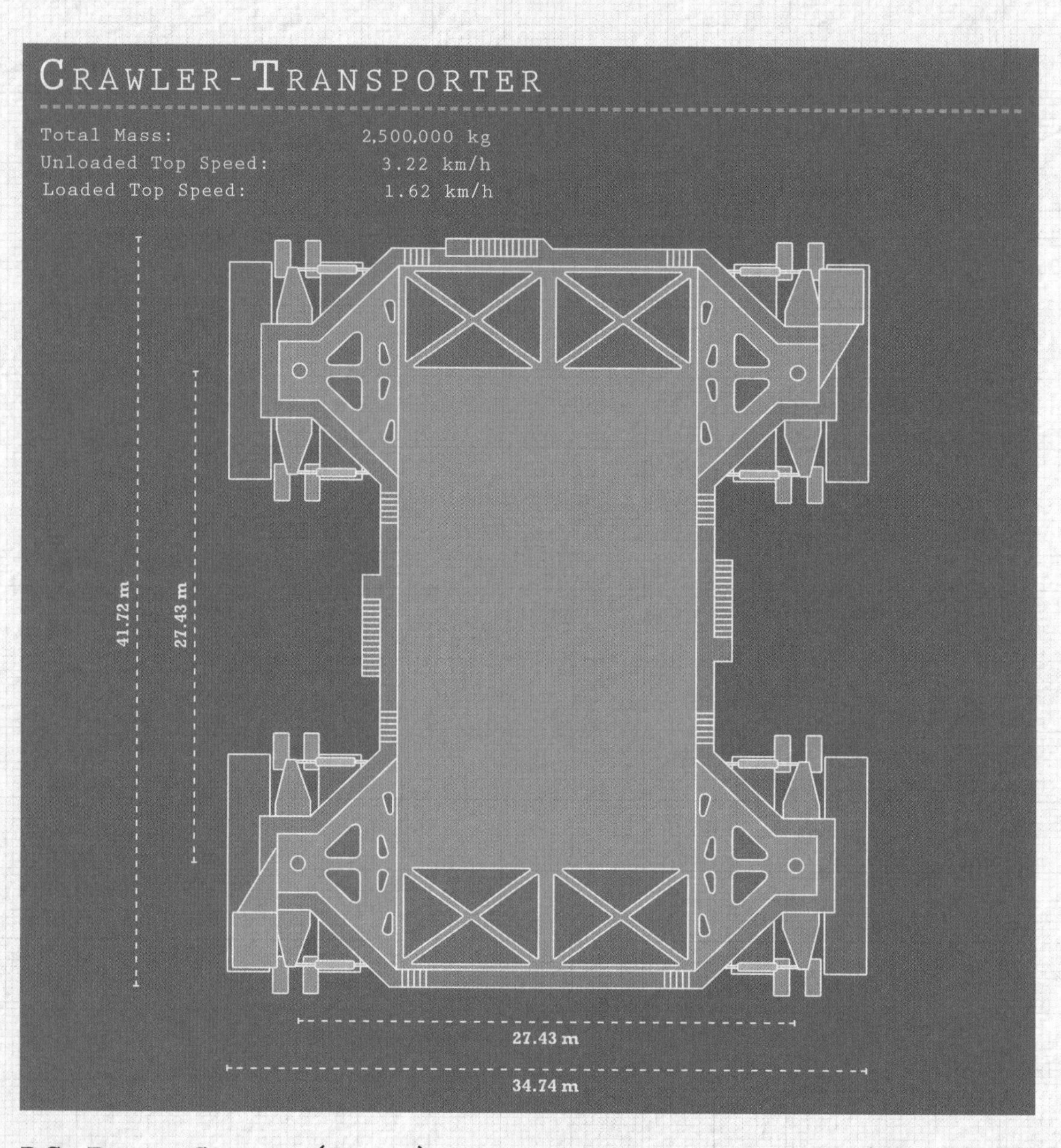

DC Power System (Drive)

Engine:	Alco 251
Generators:	2
Cylinders (each):	16
Power (each):	275 hp
Fuel:	Diesel
Generators:	4
Power (each):	1,000 kW

AC Power System (Steering)

Engine:	White-Superior
Generators:	2
Cylinders (each):	8
Power (each):	1,065 hp
Fuel:	Diesel
Generators:	2
Power (each):	750 kW

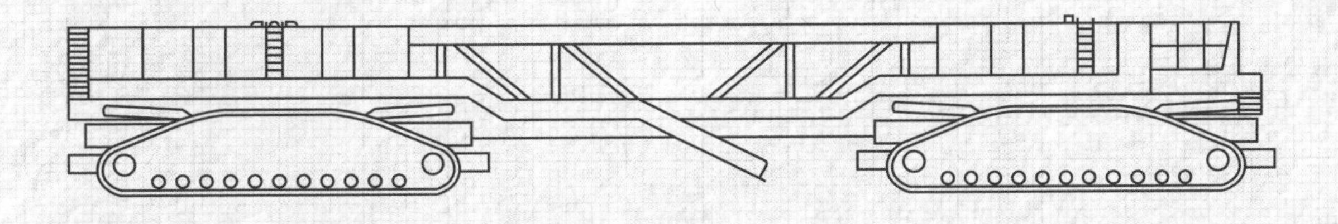

Fuel

Fuel Capacity:	19,000 L
Fuel Efficiency:	296 L/km

Loads Lifted

Chassis Only:	1,000,000 kg
MLP:	3,990,000 kg
MLP & Sat V (unloaded):	4,295,000 kg

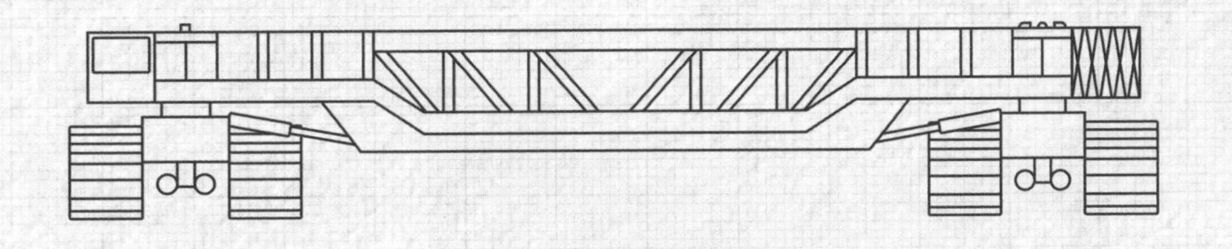

Hydraulics

Steering

Pumps:	4
Flow Rate (per pump):	2.7 L/s
System Pressure:	35,850 kPa

Jacking, Equalising & Levelling

Pumps:	8
Flow Rate (max per pump):	4.55 L/s
System Pressure:	20,700 kPa

Undercarriage

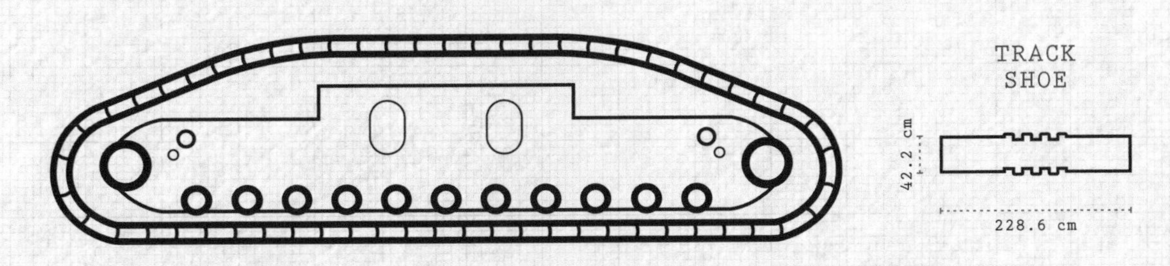

Trucks (undercarriage legs):	4
Tracks:	8
Traction Motors:	16
Traction Motor Power (each):	375 hp

Track Shoes

Quantity (per track):	57
Mass (each):	907 kg

MOBILE LAUNCHER PLATFORM MLP

The Mobile Launcher Platform (MLP) was created to support the Saturn V during assembly, transportation, and launch. It was essentially a massive steel box with a large opening that aligned with the Saturn's engines. During launch this opening would direct the blast from the rockets into the flame trench below. Three MLPs were made and, after various upgrades, are still in use with other launch vehicles. When stationary, six legs supported the MLP 6.7 m from the ground, allowing access for the Crawler-Transporter. The Launch Umbilical Tower (LUT) stood on top of the MLP. The arms that connected the LUT to the Saturn launch vehicle supplied consumables, and allowed access for servicing.

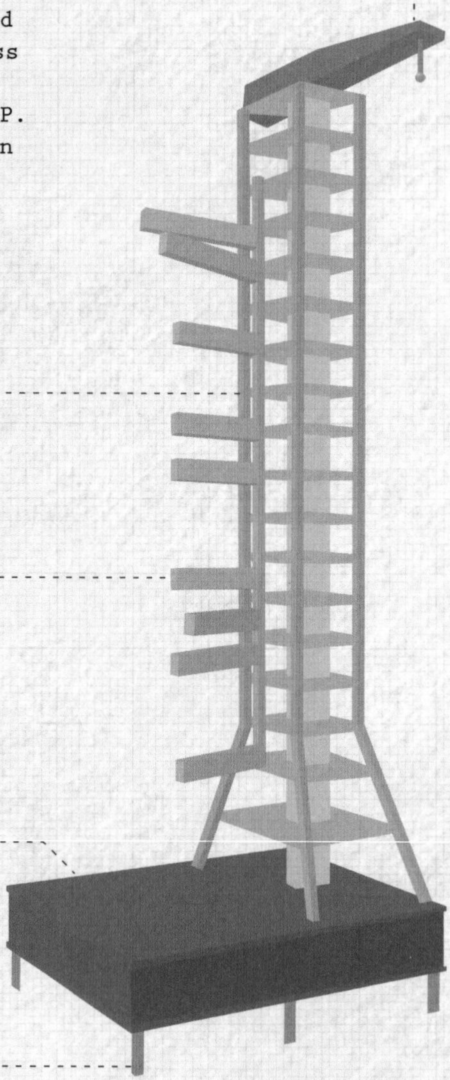

CRANE
The Hammerhead crane was capable of lifting 22.5 tonnes and could rotate through 360 degrees.

LAUNCH UMBILICAL TOWER
Standing 120 m high, it contained 18 service levels.

UMBILICAL ARMS
The 9 umbilical arms supplied the Saturn V with rocket fuel, liquid oxygen, and electrical connections. Upon launch they would release and swing out of the way.

LAUNCHER BASE
The base was split into 2 internal floors through which many heavy-duty steel girders ran.

SUPPORT LEGS

Mobile Launcher Platform (MLP)

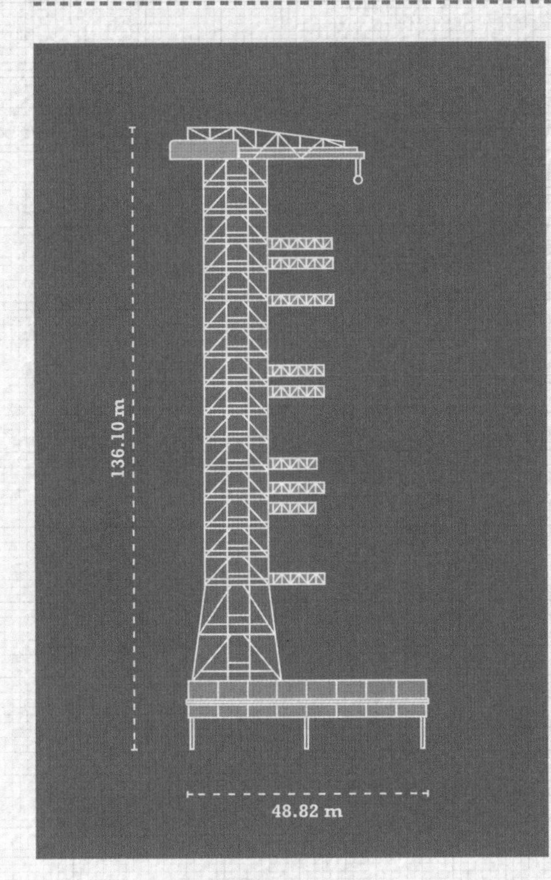

Mobile Launcher Platform

Mass:	3,730,000 kg
Levels:	2
Support Legs:	6
Level "A" Rooms:	21
Level "B" Rooms:	22

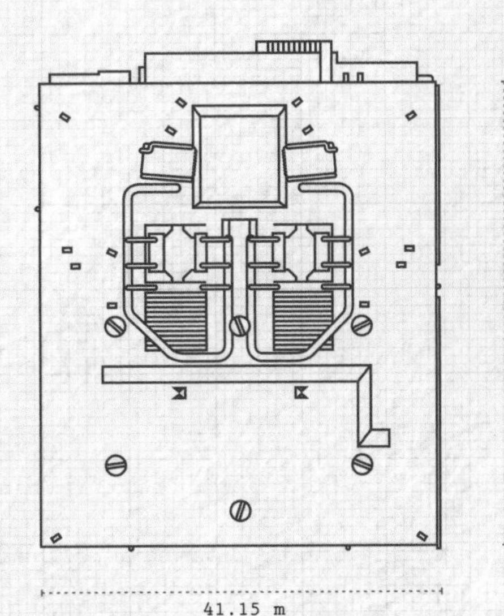

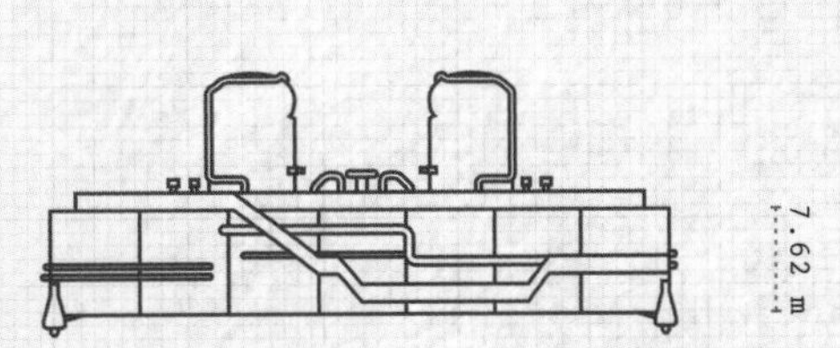

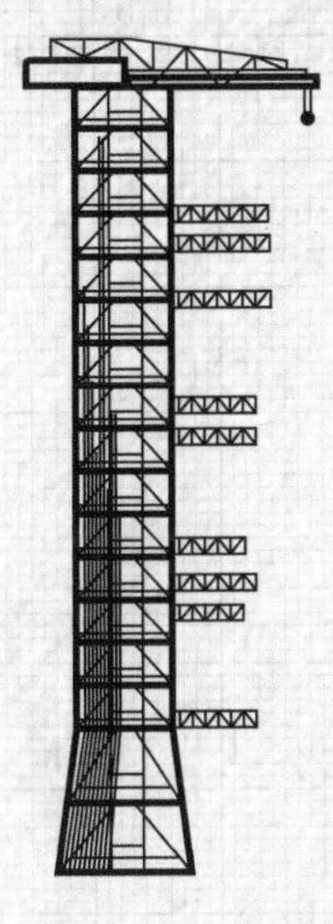

Launch Umbilical Tower

Height:	121.8 m
Mass:	1,985,000 kg
Levels:	18
Elevators:	2
Service Arms:	8
Access Arms:	1
Can Withstand Winds of:	110 km/h

VEHICLE ASSEMBLY BUILDING VAB

The Vehicle Assembly Building (VAB), or Vertical Assembly Building as it was originally known, was constructed for the vertical assembly of the Saturn V. When it was completed in 1966, it was one of the largest buildings in the world, and is still the tallest single-story building on the planet. Inside, the four bays were used for examining and stacking the rocket stages and Apollo spacecraft. A total of seventy-one cranes and hoists were installed to aid with moving the huge components. The four doors that allow access to the bays are still the biggest in the world, taking forty-five minutes to fully open or close.

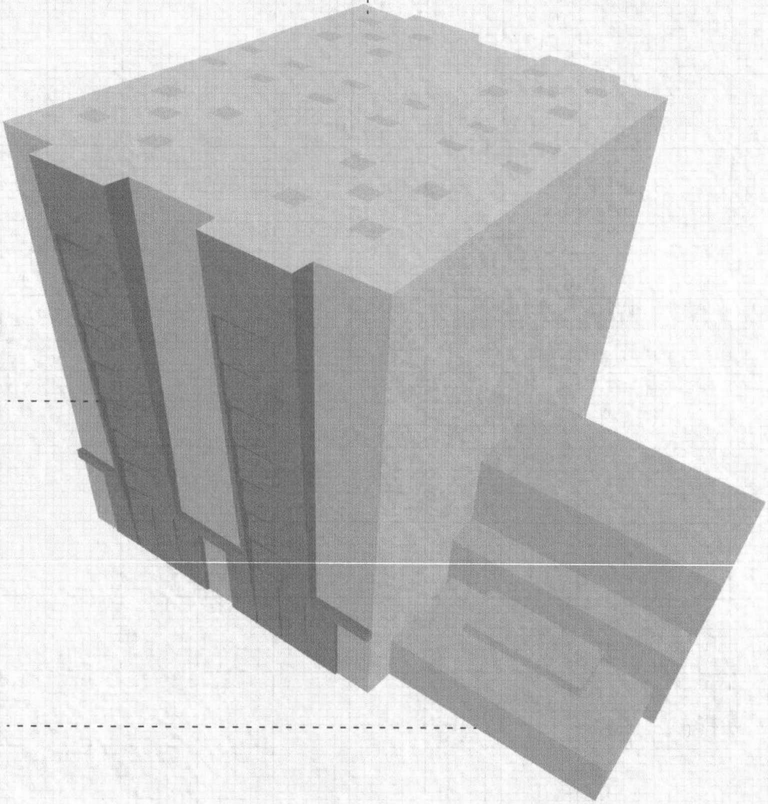

VENTILATION

There are more than 10,000 tonnes of air-conditioning equipment throughout the building, including 125 ventilators on the roof. They help to reduce humidity in the building.

BAY DOORS

Standing 139 m high, they were made from 11 sections, which were able to slide open independently.

LOW BAY

This smaller section of the building contains maintenance and overhaul workshops.

Building Plan

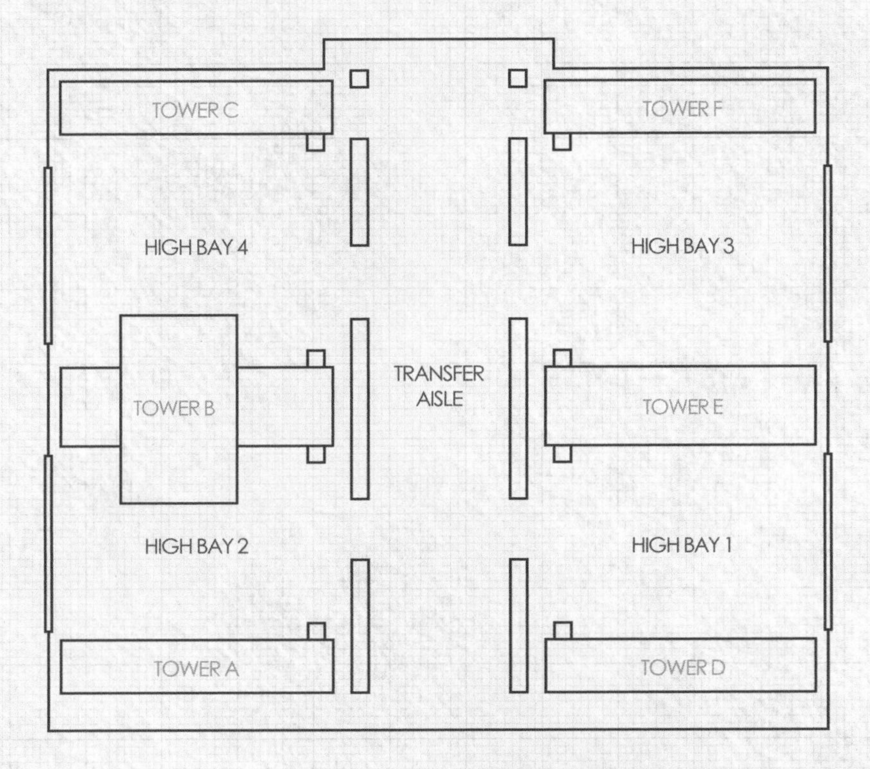

Main Doors:	4	Ground Area:	32,375 m^2
Levels:	1	Volume:	3,665,000 m^3
Lifting Devices:	71	Can Withstand Winds of:	200 km/h

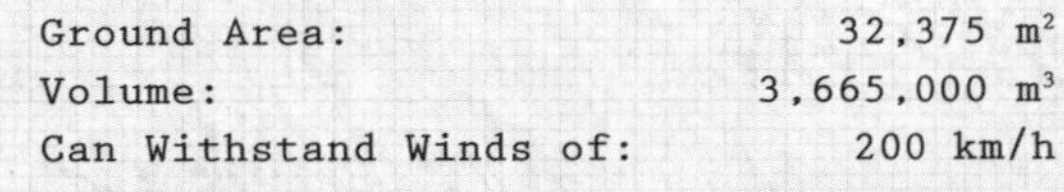

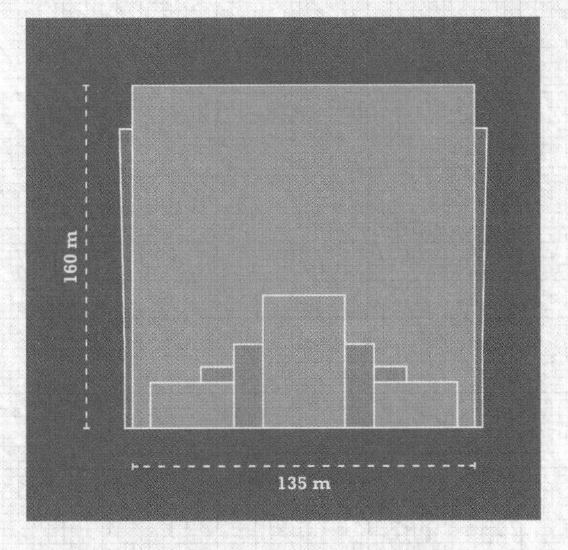

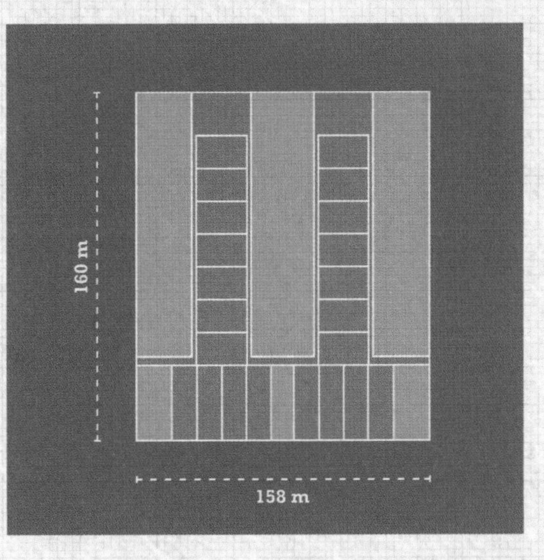

MISSIONS

Unmanned Apollo flights began in 1961 to test the Saturn launch vehicles and the space worthiness of the Apollo spacecraft. The first manned mission was in 1967, and the program achieved its goal of "landing a man on the Moon and returning him safely to Earth" before the end of 1969. By the time the program ended, NASA had performed six successful lunar landings, resulting in twelve men walking on the Moon.

In early manned missions, objectives were focused on testing the spacecraft and carrying out procedures in the orbits of both the Earth and the Moon. After Apollo 11 achieved the first landing, the mission's goals were predominantly to explore the lunar landscape, carry out scientific experiments, and collect rock samples.

Apollos 15, 16, and 17 were classified by NASA as "J-class" missions. These missions were longer, with the astronauts staying for almost three days on the Moon. The extra time allowed them to carry out more experiments and survey more of the lunar geology. The J-class missions were the only missions in which the Lunar Roving Vehicle, which extended the astronauts' range of exploration, was deployed. In addition to the J-class missions there were several other classes of mission, with letters A to I signifying operational details such as whether it was manned or not, what equipment was being tested, and what procedures would be carried out.

Over the course of the program, 382 kg of rock and lunar soil was brought back to Earth. Research of the Apollo samples has greatly contributed to our understanding of what the Moon is made from, and its geological history. Other benefits of the program include scientific advancements in telecommunications, computers, and avionics.

Plans were made for the program to run to Apollo 20, but cutbacks forced it into early retirement. After the successful landing of Apollo 11, the government reduced its funding to NASA, which at the time was also having to divert money toward the development of the upcoming Space Shuttle. In addition, one of the Saturn V rockets originally intended for the Apollo program had been reassigned to launch the Skylab space station into orbit. Due to these factors, Apollo 17 would be the last Apollo mission, and December 14, 1972 was the last time man set foot on the Moon. However, Apollo hardware would still be used in the years to come, on the historic Apollo–Soyuz mission, and ferrying crews to and from Skylab.

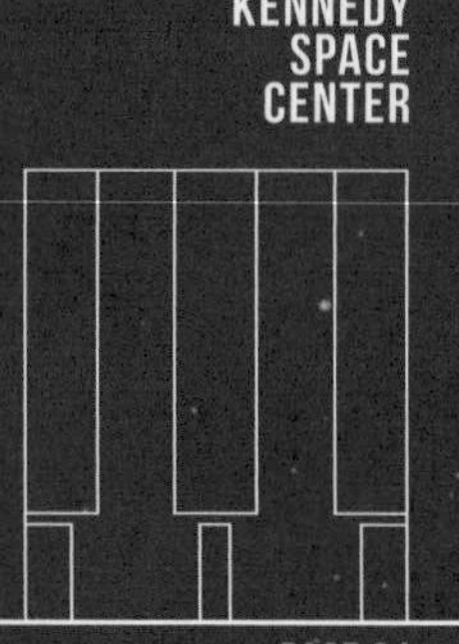

PAGE 34

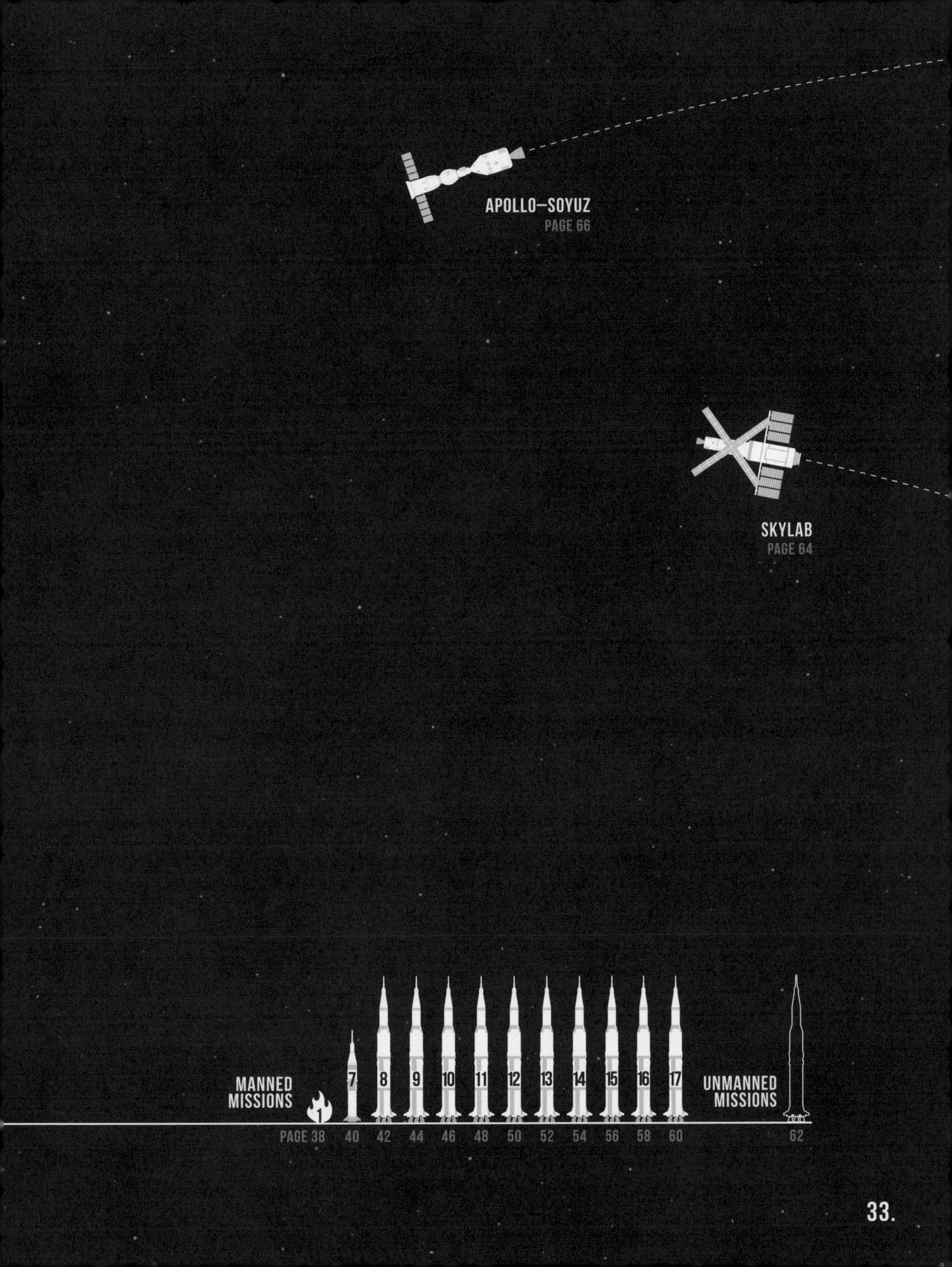
APOLLO–SOYUZ
PAGE 66
SKYLAB
PAGE 64
MANNED MISSIONS
1
7
8
9
10
11
12
13
14
15
16
17
UNMANNED MISSIONS
PAGE 38
40
42
44
46
48
50
52
54
56
58
60
62

KENNEDY SPACE CENTER

The US military had been conducting rocket launches from Cape Canaveral Air Force Station, on the east coast of Florida, since 1949. Its location was chosen because it allowed for launches over the ocean, which minimized risk to the population, and its proximity to the equator gave the rockets an extra boost from the Earth's rotation. Project Mercury, Project Gemini, and a few early Apollo missions were all conducted from here; however, the facility was deemed unsuitable to accommodate the Saturn V, and so agency officials selected another site.

Construction began on the Kennedy Space Center in November 1962 on Merritt Island, in close proximity to the existing air force station. Here they would construct the towering Vehicle Assembly Building, with its adjoining Launch Control Center; 40 meter-wide crawlerways; the Operations and Checkout Building for astronaut and crew accommodation; a press site; a whole host of support buildings; and a brand-new launch complex. This would feature two pads, 39A and 39B (the smaller 39C was added in 2015), and was safely situated nearly 5 km from the Launch Control Center. During launch preparations the astronauts would be linked by radio with the ground team at the Launch Control Center, and only after launch, when the rocket had cleared the tower, would communications switch to CAPCOM in Houston, Texas. As the climate in Texas is drier and less prone to storms, it was a more suitable place to base their center for communications.

The first launch from the Kennedy Space Center was an unmanned mission, Apollo 4 in 1967, which lifted off from 39A. All manned missions would lift off from this pad with the exception of Apollo 10, which used 39B because Apollo 11 (the first Moon landing) was already being prepared on 39A. Following the Apollo, Apollo–Soyuz, and Skylab missions, the Kennedy Space Center was modified for the Space Shuttle program, which ran until 2011.

SpaceX, a privately owned spaceflight company, is currently operating from pad 39A. After signing a twenty-year lease of the property in 2014, they have been overhauling the premises to make them suitable for their Falcon rockets, including modifications to the pad and the construction of a new assembly building. Meanwhile, since 2012, NASA has been preparing pad 39B for the new Space Launch System, a rocket capable of similar thrust to that of the Saturn V. NASA will also be making this pad available to other commercial users when not needed by the Space Launch System. The newest pad, 39C, was built within 39B's perimeter. It was designed to accommodate more diminutive rockets, in the hope that smaller companies might have a chance to break into the commercial spaceflight market.

CAPE CANAVERAL LAUNCH COMPLEXES

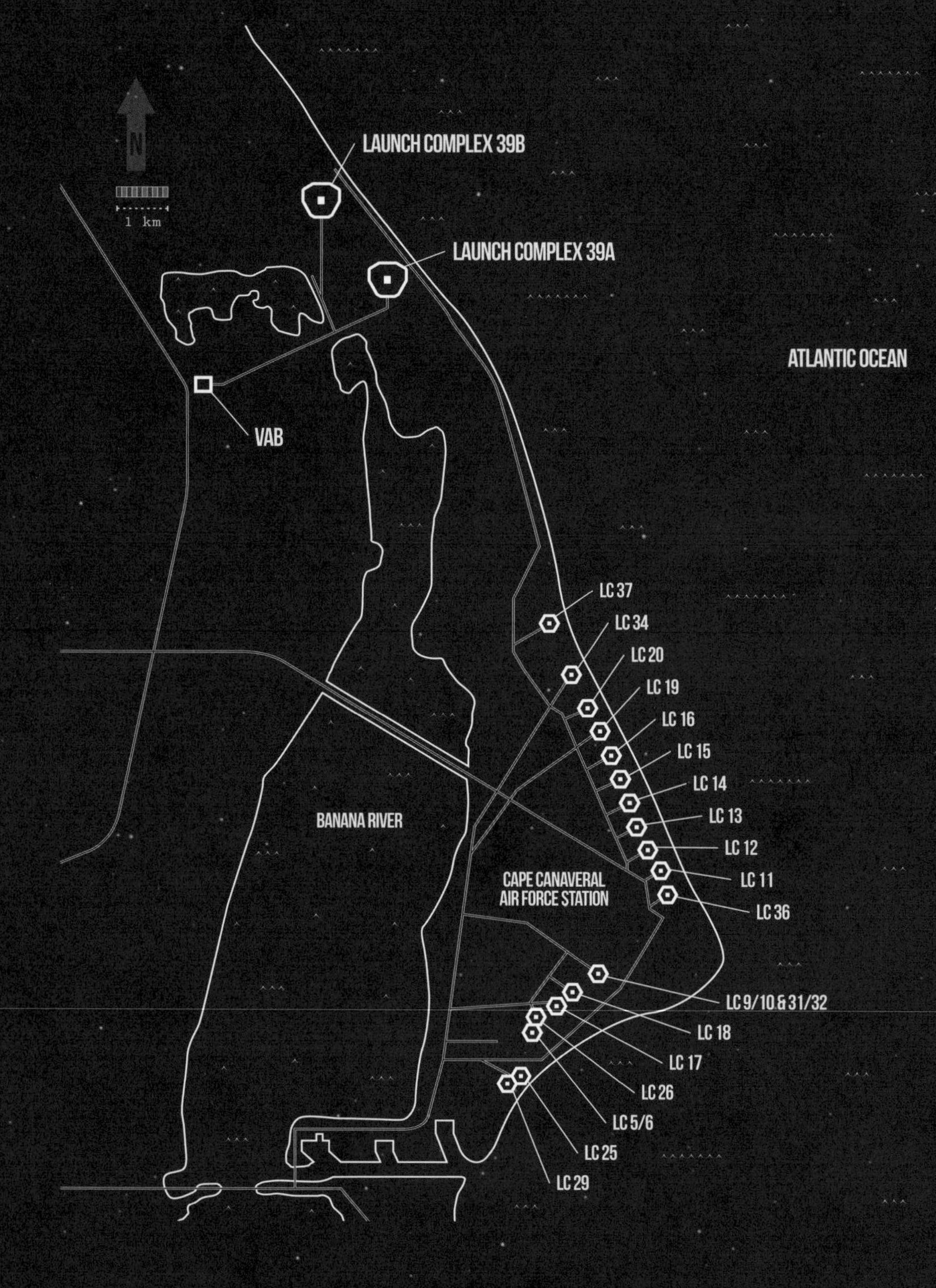

LAUNCH COMPLEXES 39A & 39B

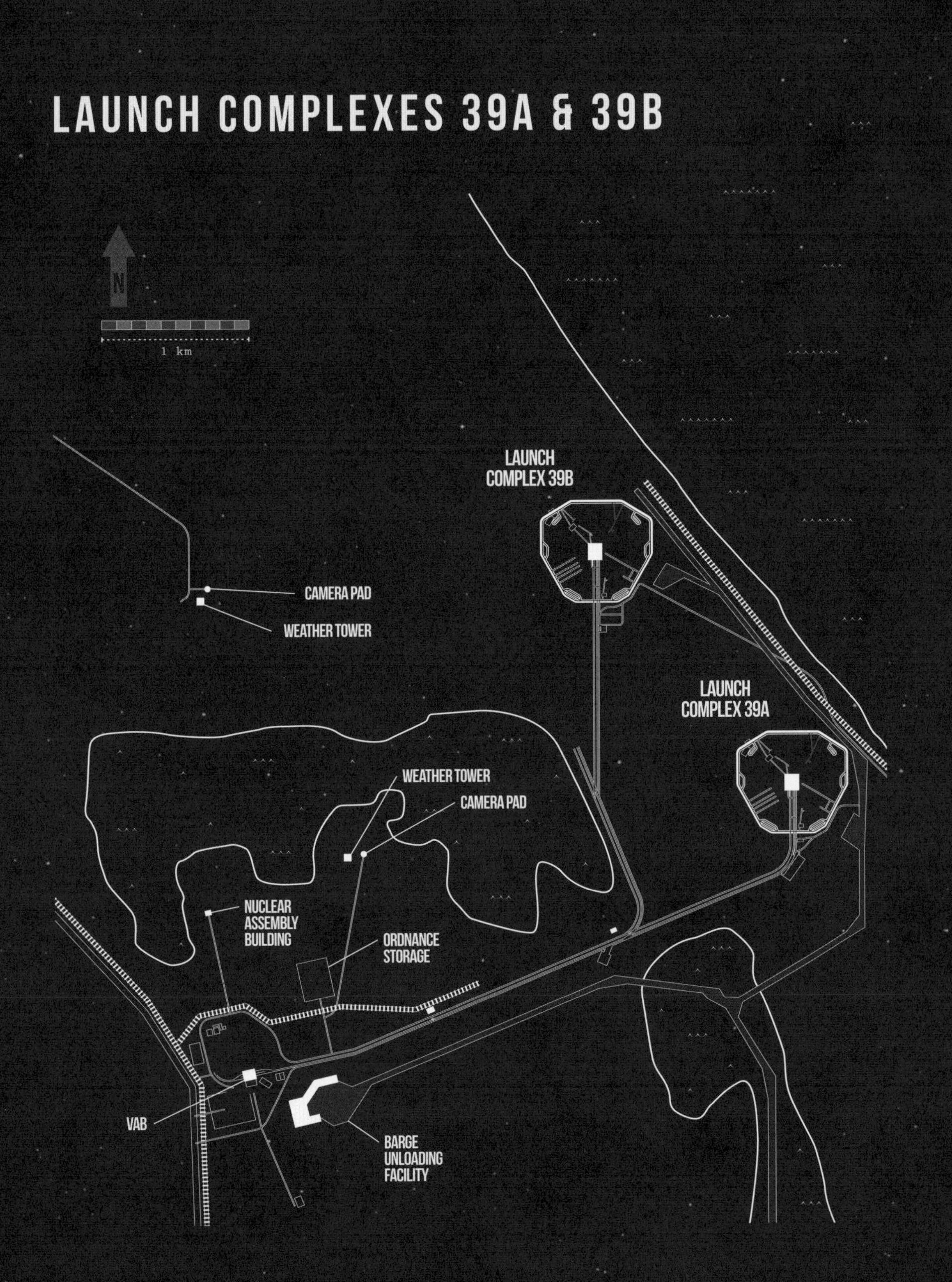

APOLLO 1

The first manned mission of the Apollo program, Apollo 1, originally designated AS-204, had been due to launch on February 21,1967, but the mission was cancelled after a disaster during a preflight test. Three astronauts, Virgil Grissom, Edward White, and Roger B. Chaffee, lost their lives. It was in honor of this tragedy, and by the wishes of the crew's wives, that their test mission was renamed Apollo 1.

The incident took place during a launch rehearsal test at Cape Canaveral Air Force Station, Florida, on January 27, 1967. The three crewmen were seated in launch positions, sealed inside the Command Module, when fire swept through it. The pressurized, oxygen-rich environment helped the blaze to spread quickly. Ground crew rushed to their aid, but before they could reach them the CM ruptured, billowing out flames and thick black smoke.

Afterward, NASA convened an accident review board to find the cause of the fire. Although the exact source of ignition was never determined, the board discovered a wide range of design and construction flaws. While the major engineering modifications were carried out, manned Apollo flights were suspended. The next mission to carry humans on board would be in twenty months' time: Apollo 7.

APOLLO 7

Apollo 7 launched from Cape Kennedy Air Force Station at 11:02 EST on October 11, 1968. The crew, Commander Walter M. Schirra, CM Pilot Donn F. Eisele, and LM Pilot Walter Cunningham, was aboard the only manned Apollo mission to lift off using the Saturn IB launch vehicle. It was also the last manned mission from Cape Kennedy Air Force Station.

Apollo 7's objective was the same as that of the ill-fated Apollo 1: an Earth orbital test flight of the Command/Service Module with a crew on board. Since the mission was to test the CSM, there was no Lunar Module aboard. The reduced weight as well as the craft only travelling into low-Earth orbit were the reasons for using the Saturn 1B; all future missions would utilize the more powerful Saturn V.

The flight was a total success, with the craft and crew orbiting the Earth 163 times in just under eleven days. During their time inside the CM, the astronauts broadcast the first live television from space. The CM was a lot more spacious than on the previous American space programs (Mercury and Gemini), and the availability of hot meals made their stay in orbit more tolerable. However, despite these efforts to make their environment more comfortable, the prolonged journey took its toll on the crew. The astronauts became irritable, complaining about the selection of food (predominantly the high-energy sweets) and talking back to CAPCOM. The situation was compounded when Schirra got a cold. However, these were only minor troubles on an otherwise tremendously successful mission. On the tenth day, 259:43 Ground Elapsed Time (GET), the CM separated from the SM to prepare for reentry. The crew would position the CM to enter Earth's atmosphere blunt-end first, in order to create maximum air resistance, the most effective way of slowing the craft. Further braking force was provided by the CM's parachutes, and the three men splashed down in the Atlantic on October 22, 1968 (260:09 GET). The mission's success gave NASA the confidence they needed to launch Apollo 8 two months later with the intention of orbiting the Moon.

APOLLO 8

This was the second manned flight in the Apollo program. The objective was to send the Apollo CSM to orbit the Moon and return its crew to Earth.

Commander Frank Borman, CM Pilot James A. Lovell, and LM Pilot William A. Anders were the crew at the forefront of another blinding success that achieved many firsts: the first manned spacecraft to leave the Earth's orbit, the first manned spacecraft to reach and orbit the Moon, the first live television pictures of the Moon's surface, the first manned launch of the Saturn V, and, in a breathtaking moment, the crew members were the first humans to see the Earth as a whole planet. Most importantly, though, having confirmed that the CM was worthy of spaceflight in Apollo 7, NASA had now proved that their flight trajectory and operations for getting to the Moon and back were sound.

Apollo 8 launched on December 21, 1968, at 07:51 EST from Cape Kennedy Space Center, the site from which all future Apollo missions would lift off. Upon launch the first two stages of the Saturn V rocket were used and jettisoned in turn, accelerating the craft through the upper atmosphere. The third stage then fired, carrying the astronauts to just over 28,000 km/h, placing them in Earth's orbit. In orbit the crew performed system tests and readied the craft for trans-lunar injection (TLI), which occurred at 02:50 GET. This is when the third stage of the Saturn V fired for the second time, propelling them out of the Earth's orbit and on course for the Moon. As there were no plans for a lunar landing, there was no Lunar Module aboard the spacecraft. Instead a "Lunar Test Article" sat in its place, which was equivalent in weight and would act as ballast.

The Apollo spacecraft took three days to travel to the Moon, at which point the Service Propulsion System (SPS) engine was employed to slow them down, placing them in an elliptical orbit. Here the SPS engine fired again, and within two revolutions they were placed in a near circular orbit around 110 km from the Moon's surface. During the mission the crew made a live television transmission on Christmas Eve, broadcast to the largest ever worldwide audience at the time. The Apollo 8 spacecraft travelled a total of ten times around the Moon in roughly twenty hours, while the astronauts surveyed and photographed the lunar surface. On their final orbit, while they were on the far side of the Moon, and once the craft was readied for trans-Earth injection (TEI), they fired the SPS engine again. This gave the CSM the speed it needed to leave the Moon's orbit and send them on a trajectory back to Earth. The return journey provided an opportunity for the astronauts to relax while still monitoring the spacecraft's instruments. Shortly before reentry they jettisoned the SM. The CM and its crew returned on December 27, 146:59 GET, splashing down in the North Pacific.

APOLLO 9

Apollo 9's primary aims were to test the LM as a self-sufficient spacecraft, and to practice the rendezvous and docking procedures between the LM and the CSM. The mission launched at 11:00 EST on March 3,1969. As the LM was now on board, this would be the first time that the Apollo/Saturn V spacecraft would lift off in full lunar-landing mission configuration. Its crew members were Commander James R. McDivitt, CM Pilot David R. Scott, and LM Pilot Russell L. Schweickart.

Apollo 9 would carry out its objectives in Earth's orbit, while the LM and CSM would dock with each other twice. The first took place on day one, after the Saturn V's third stage (the S-IVB) had placed the craft in Earth's orbit. With the LM still attached to the third stage, the CSM would separate, turn 180 degrees, and then align its docking hatch with that of the LM. The CSM then gently thrusted to make the connection and lock the two craft together. The third stage was subsequently detached from the Apollo vehicles, and, using one final thrust, it was sent on a trajectory that would take it to orbit the Sun.

A few days later, after functional tests had been carried out in the LM, Schweickart readied himself for an EVA. Wearing the newly designed Apollo spacesuit, he climbed out of the LM's porch. This suit, which contained its own life-support mechanisms, was the first of its kind; until then astronauts relied on an "umbilical" to connect them to their spacecraft. Schweickart's space walk took just over thirty-seven minutes; although he had been due to stay out longer, sickness caused some of his activities to be scrapped. Scott captured the spacewalk on film from the CM.

On the fifth day of the mission, McDivitt and Schweickart boarded the LM so that they could rehearse some of the procedures the LM would carry out during a lunar landing. This involved separating themselves from Scott in the CM, and instead of performing a landing, they used the LM's descent-stage engine to thrust themselves into a higher orbit 179 km away. With the two craft reunited after six hours apart, the astronauts transferred over to the CSM and jettisoned the LM, where it would remain orbiting the Earth until it decayed.

The mission had proved that the LM was capable of spaceflight and that it could perform its necessary tasks. Splashdown occurred on March 13, 1969 (241:00 GET), in the North Atlantic, with the crew and craft having completed 152 revolutions of Earth.

APOLLO 10

Lift-off took place at 12:49 EST on May 18, 1969. It was to be a rehearsal of the Moon landing planned for Apollo 11, although this time they would not be touching down on the Moon. The astronauts aboard were Commander Thomas P. Stafford, CM Pilot John W. Young, and LM Pilot Eugene A. Cernan.

As the Apollo vehicle began its orbit of the Earth, the astronauts prepared for TLI. This occurred after one and a half revolutions when the Saturn V's third stage reignited, increasing their speed and propelling them toward the Moon. Shortly after the third-stage rocket had been used the crew would carry out the first of the docking procedures that had been practiced in Apollo 9; however, this time they would not be in Earth's orbit. As they made their way through space, the CSM detached from the third stage, then turned around and docked with the LM. The last stage of the Saturn V launch vehicle then separated from the Apollo CSM/LM, and they continued on toward the Moon.

Three days after lift-off, with a burn of the SPS engine to slow them down, they arrived at the Moon, entering into an elliptical orbit. As before (in Apollo 8) the SPS would fire again to place them in a circular orbit roughly 110 km from the surface. On the fifth day, Stafford and Cernan boarded the LM and, leaving Young alone in the CM, descended toward the Moon. The LM would reach an altitude of just 14.3 km, where it would remain in orbit until it was time to return to the CSM.

As they swooped closer to the Moon than any human being had been before, twice passing over the Sea of Tranquillity (the intended Apollo 11 landing site), Stafford and Cernan took as many photographs as they were able. They were so close that there was genuine concern that they might be tempted to land and thus become the first men on the Moon. In order to prevent this, the LM's ascent stage was only partially fuelled. Had they decided to bring it down and go for a landing, they would have been stranded, as the ascent stage would not have had the fuel to return them all the way to the orbiting CSM.

Once it was time for the LM to return to the CSM, the LM's descent stage was jettisoned and left in lunar orbit, eventually to be pulled in by the Moon's gravity to crash into the surface. The ascent-stage engine fired them back toward the orbit of the CSM, and the two craft docked back with each other approximately eight hours after their separation. With all the crew members back aboard the CSM, the remainder of the LM was jettisoned and the crew began their journey back to Earth. During their return, Apollo 10 set the record for the highest speed attained by a manned vehicle: 39,897 km/h. This figure wasn't beaten by any of the subsequent Apollo missions and is still held to this day. The crew reentered and landed the CM in the Pacific on May 26, 1969 (192:03 GET).

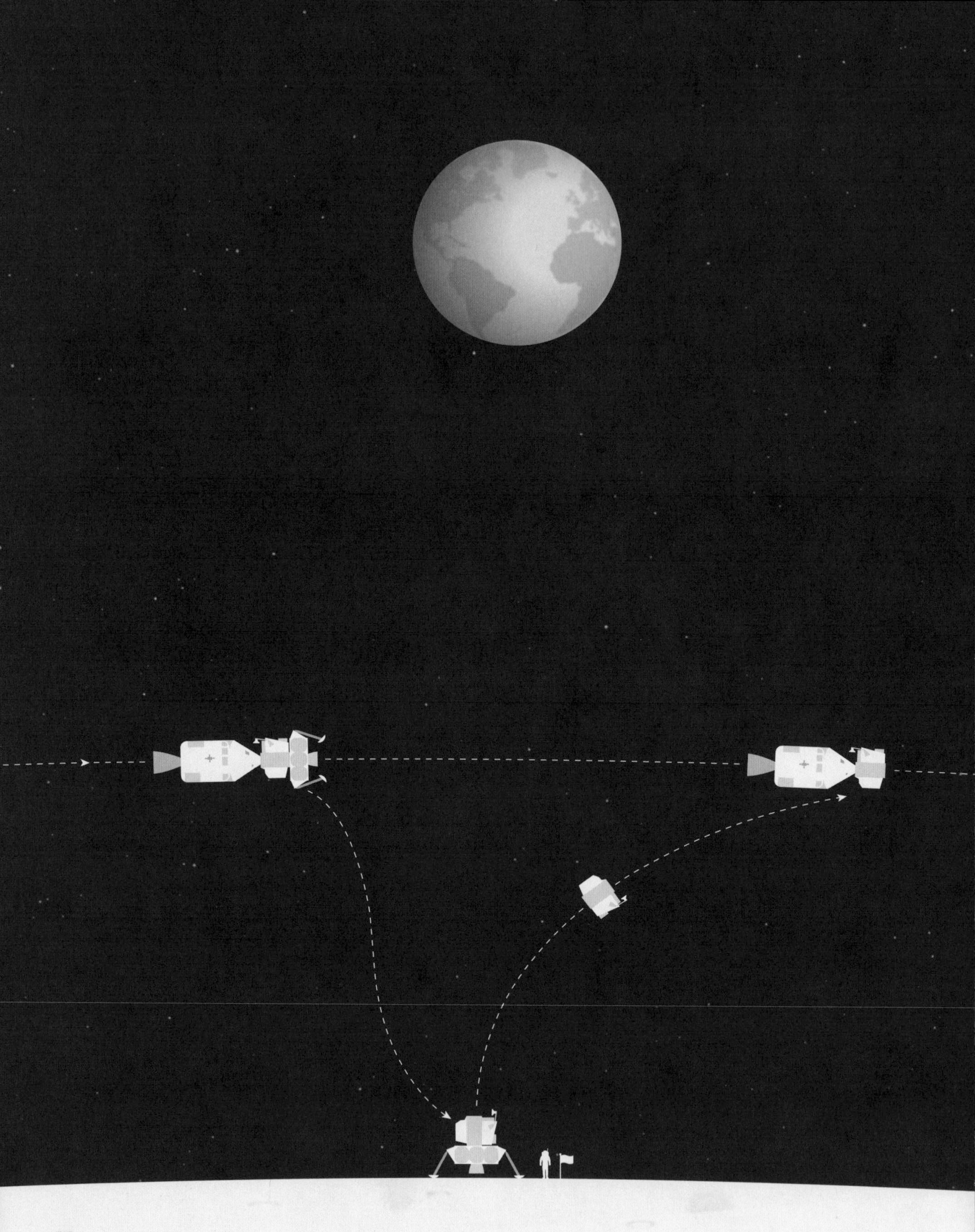

APOLLO 11

Apollo 11 was the fifth manned Apollo mission and marked the first time man set foot on another celestial body. Astronauts Commander Neil Armstrong, CM Pilot Michael Collins, and LM Pilot Edwin "Buzz" Aldrin launched at 13:32 EST on July 16, 1969, from the Cape Kennedy Space Center.

Similar to previous lunar excursions, the Apollo craft would spend two and a half hours in Earth orbit (completing roughly one and a half revolutions) before TLI, which fired them free of Earth's gravity and toward the Moon. After roughly three days travelling, they arrived at the Moon, entering into an elliptical orbit at 75:50 GET. As rehearsed, the SPS engines fired in order to bring the craft into a circular orbit 110 km in altitude. Twenty-five hours into their lunar orbit, they were ready to undock; Collins was to stay in the CM while Armstrong and Aldrin separated from him in the LM. Following a thorough check of all the LM systems, the descent-stage engines were fired for twenty-nine seconds, and they began their course for the lunar surface. Almost an hour later, the descent stage was fired again for the final time. This was in order to slow the LM's approach and give Armstrong enough time to navigate them away from a boulder-strewn area he could see the craft was heading toward. They touched down on the Moon on July 20, 1969, 102:45 GET.

Immediately after landing, procedures required them to prepare the LM for launch, as a contingency measure. After these preparations it was Armstrong who emerged from the vehicle first. While descending the ladder he had to activate a television camera on the side of the LM to document this momentous occasion. Becoming the first human to place his foot on another world, he famously proclaimed, "That's one small step for a man, one giant leap for mankind." Aldrin joined him moments later, describing what lay before him as "magnificent desolation." The two of them spent two and a half hours on the surface collecting samples of lunar rock and taking photographs. They also deployed some scientific equipment, including a Laser Ranging Retro Reflector, for the reflecting of lasers beamed from Earth. Before they left they erected the US flag and left a plaque commemorating the momentous occasion. After boarding the LM and sealing the porch hatch, the ascent stage was uncoupled from the descent stage. The two astronauts then launched from the surface of the Moon to rejoin Collins, who remained in orbit. The LM successfully located the CSM and, after some careful positioning, the two craft docked together, allowing the crew (and their lunar rocks) to transfer over to the CSM for the journey home. The crew returned triumphantly with a splashdown in the Pacific on July 24, 1969, 195:18 GET.

APOLLO 12

This was the second mission to land man on the Moon. It launched four months after Apollo 11, at 11:22 EST on November 14, 1969, with a crew of Commander Charles Conrad Jr., CM Pilot Richard F. Gordon Jr., and LM Pilot Alan L. Bean. They were aiming to make a precision landing near the site of Surveyor 3, an unmanned probe that the United States had landed on the Moon on April 20, 1967.

A harrowing moment came when lightning struck the spacecraft just thirty-six seconds after lift-off. In the CSM warning lights lit up everywhere, various systems automatically shut down, and the crew in the air and on the ground frantically tried to assess the situation. John Aaron, a quick-thinking systems engineer at Mission Control, remembered a similar pattern of failings from an earlier test and instructed the crew how to reboot the power system. With telemetry systems re-established and the CSM power restored, launch proceeded successfully, and the remainder of their voyage went as planned. On their arrival at the Moon, Conrad and Bean boarded the LM and separated from Gordon, who was left to orbit in the CM. The descent of the LM was governed automatically, with Bean making only minor manual corrections during the final phases. Touchdown was an enormous success; the LM landed only 183 m away from the target spacecraft, in an area known as the Ocean of Storms.

Not long after alighting from their craft, television transmission was lost, as Bean had damaged the sensitive camera by inadvertently pointing it at the Sun. The pair spent thirty-one and a half hours on the lunar surface, during which time they performed two EVAs, with a seven-hour rest period in the LM in between. On their excursions they visited the Surveyor 3 probe to remove some of its components so they could later be analyzed. The results of these examinations yielded important information on the effects that long-term exposure to the lunar environment has on materials. Another task that the astronauts had to carry out was a geology traverse. Conrad and Bean covered approximately 1,300 m on foot, taking regular samples while documenting their findings.

The astronauts also set up and left behind a piece of equipment that would measure solar winds, magnetic fields, and the Moon's seismic activity, among other things. It was known as the Apollo Lunar Surface Experiments Package (ALSEP) and, as it had its own power supply, it could relay data back to Earth over a long period of time. Once the two astronauts left the Moon and returned to the CSM, the jettisoned LM was sent to crash into the Moon to provide a controlled impact for ALSEP to measure. The mission ended on November 24 with a splashdown in the Pacific, 244:36 GET.

APOLLO 13

Apollo 13 launched on April 11, at 14:13 EST. On board were Commander Jim Lovell, CM Pilot Jack Swigert, and LM Pilot Fred Haise. Their destination was an area of the Moon called Fra Mauro.

The astronauts were nearly fifty-six hours into the mission (about 330,000 km from Earth) when they heard a loud bang come from behind them, somewhere in the SM. The crew briefly lost communications with CAPCOM because of damage to the high-gain antenna, although this swiftly self-rectified by switching bands. Outside the CM window the men could see fluids of some kind being ejected from their craft into space. The crew stayed calm and liaised with Mission Control, reporting what they were witnessing. Readings confirmed that the bang they heard was one of the SM's oxygen tanks exploding, a very serious problem for those on board. They relied on this oxygen not only for breathing, but for producing water and electrical power. It was clear very quickly that there would be no Moon landing. This time, the mission was to survive. Mission Control decided the best course of action for Apollo 13 was to bring it home the fastest way possible. Their plan was to continue on toward the Moon and enter its orbit for half a revolution. Before the half-orbit was complete, the SPS engines would fire up again to generate the speed needed to break free from the Moon's gravity and send them on a return course to Earth.

Due to the damage sustained, the CSM was powered down at 58:40 GET and the crew boarded the LM, which was configured to supply all the necessary power and other consumables. The LM, however, was not designed to support three people for such a length of time, and CO_2 levels were rising fast. To avoid the crew being poisoned by their atmosphere, NASA set its brightest minds to work on a problem that was worsening with every breath the astronauts took. Ingeniously, NASA's team on the ground managed to craft a working component out of the odds and ends that they knew the crew had on hand. It was a device that allowed the astronauts to adapt the CO_2 filters from the CM to work with the LM's system. CAPCOM communicated the instructions to the crew and they immediately set to work. As soon as they'd implemented the makeshift repair, to their relief, CO_2 levels began to fall and returned to breathable levels. Another hardship endured by the crew was the plummeting temperature. The need to conserve what little power the craft had meant that heating was kept to an absolute minimum. At times it got so cold that water would condense on the interior surfaces, causing further concerns that the electronics would short.

Apollo 13's new trajectory succeeded in guiding them safely around the Moon, and they thrust out of its orbit as planned. Although in relative discomfort, they at least knew that they were on their way home. Before entering Earth's atmosphere the crew reentered and powered up the CM. They then jettisoned the blighted SM, before jettisoning their savior, the LM, which had been their lifeboat for the journey. The crew members, who were in generally good condition, were picked up in the South Pacific by recovery ship USS *Iwo Jima*, after their splashdown on April 17, 1970, 142:54 GET.

APOLLO 14

Apollo 14's objectives were a reprise of the previous, aborted mission. And like Apollo 12 its aims were for the commander (Alan Shepard) and LM pilot (Edgar Mitchell) to spend more than thirty hours on the Moon and carry out two EVAs, while the CM pilot (Stuart Roosa) would remain in lunar orbit. Although the mission was ultimately successful, it was beset with problems.

Launch was delayed by bad weather conditions at the launch site—a first for the Apollo missions. Nevertheless circumstances improved and lift-off took place at 16:03 p.m. EST on January 31, 1971. Just after trans-lunar injection, the crew encountered complications with the first docking procedure. It took them an hour and forty-two minutes to capture and dock with the LM, and they even had to resort to using the SPS engine rather heavily to thrust the two craft together.

On the LM's descent to the Moon, the crew began receiving "abort" signals due to a faulty switch. If this was left unchecked, there was a risk the onboard computer might override the descent, jettison the descent stage, and power back into orbit using the ascent engine. NASA's software teams hurriedly found a solution and communicated the instructions to Mitchell aboard the LM. He managed to manually enter the modification into the LM's computer, overriding the faulty signal just in the nick of time. As their approach grew near, the LM's radar began to glitch, hampering the crew's ability to judge ground speed and altitude. The system was rebooted, and they regained their readings at 5,500 m from their destination. Despite this, the craft landed just 53 m from its target, the most accurate LM landing of all the Apollo missions.

During the EVAs the astronauts would again deploy scientific equipment, explore the landscape, and bring back samples. To help them haul their materials around, they had with them the Modularized Equipment Transporter (MET), a two-wheeled trolley that also acted as a portable workbench. This was the first and only use of the MET. As with Apollo 12, another ALSEP was deployed to send more data about the Moon back to Earth. One of the more memorable moments of the mission came toward the end of their stay on the lunar surface, when Shepard produced a golf club that he'd brought with him. Due to the limited mobility of his spacesuit, he had to swing one-handed, but managed to connect well on his second attempt. The ball flew for hundreds of meters due to the low lunar gravity, equivalent to one sixth of the Earth's. At the end of their stay, Shepard and Bean boarded the LM and prepared it for their return to the CSM. They launched from the Moon thirty-three and a half hours after their arrival to locate and dock with Stuart Roosa, who had been carrying out other experiments and taking high-resolution photos of future landing sites while in orbit. Once the crew was all aboard the CM, the LM was again sent to crash into the Moon, where its impact would be measured (by two ALSEPs this time). They returned to Earth just under three days later with a landing in the Pacific on February 9, 1971, 216:01 GET.

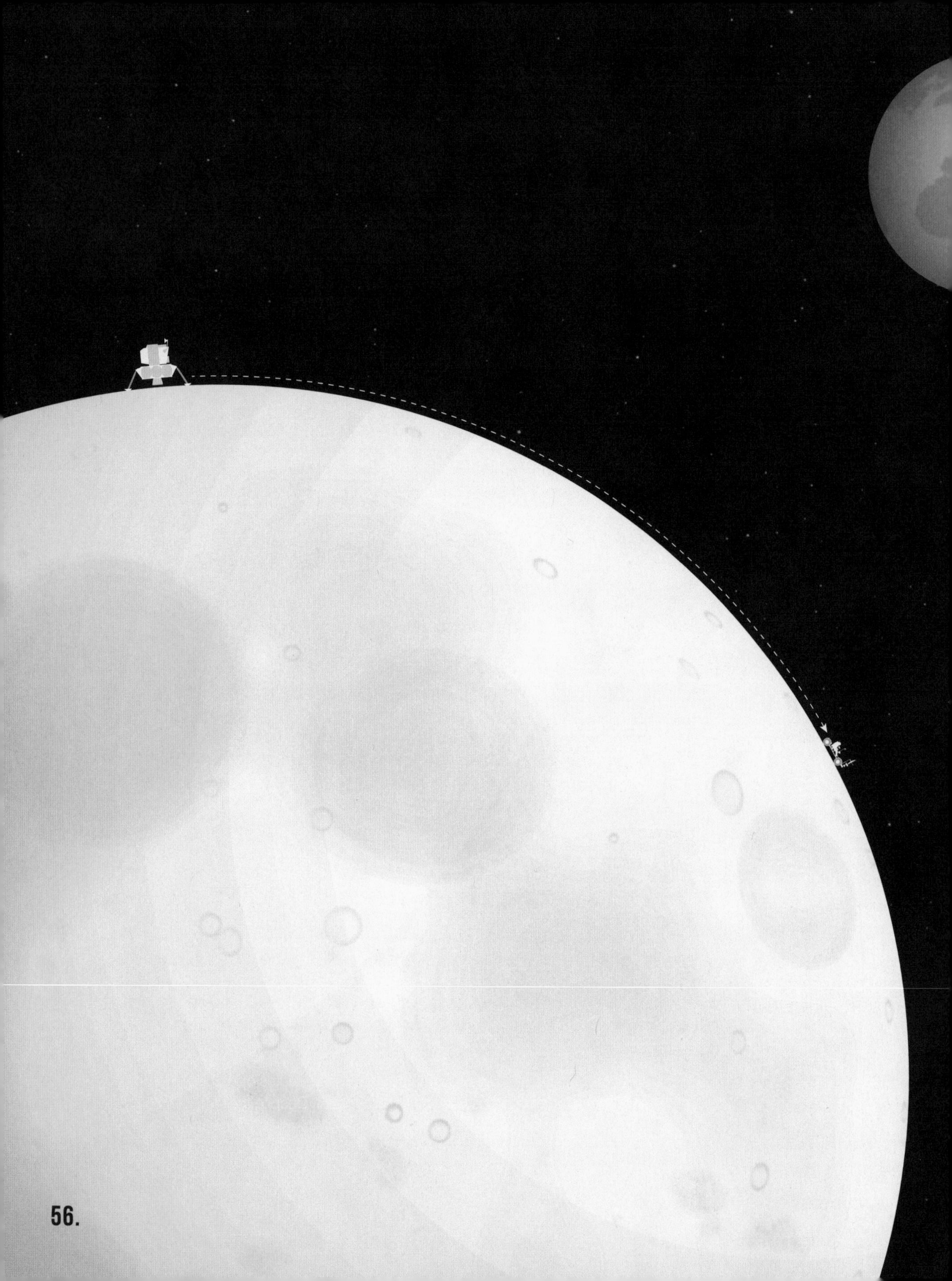

APOLLO 15

This mission was launched on schedule at 09:34 EST, on July 26, 1971. It was the first mission with an extended stay on the Moon; the two astronauts aboard the LM—Commander David R. Scott and LM Pilot James B. Irwin—would be on the surface for almost three days.

Having the Lunar Roving Vehicle with them, coupled with their prolonged stay, allowed the astronauts to travel further from their landing site than on earlier missions. Previously, surface traverses would total in the region of hundreds of meters; now they could roam tens of kilometres away from home. Exploring a much wider terrain, they examined the geology and took samples, bringing back a much greater haul than had been possible before. The two astronauts also set up another ALSEP, the third in operation on the Moon at this point.

While the two astronauts carried out their tasks on the surface, CM Pilot Alfred Worden was busy in orbit. Using a variety of cameras and mapping devices housed in the Scientific Instrument Module (SIM) in the SM, he studied the lunar surface and environment. At times Worden would be 3,600 km away from his comrades on the lunar surface; no human being in history has ever been so isolated. It would change his perspective on the universe forever: "What I found was that the number of stars was just so immense. In fact I couldn't pick up individual stars, it was like a sheet of light. . . . There are billions of stars out there—the Milky Way galaxy that we're in contains billions of stars, not just a few. And there are billions of galaxies out there. So what does that tell you about the Universe? That tells you we just don't think big enough. To my mind that's the whole purpose of the space program, to figure out what that's all about."

After all the Apollo 15 crew members had been reunited, they released a small subsatellite, PFS-1, which was to orbit the Moon, measuring its plasma, particle, and magnetic field environment. It would remain in the Moon's orbit, returning data, until January 1973. The next day, during their return journey, Worden performed the first spacewalk of its kind—in deep space. The purpose of this was to retrieve used films from the SIM bay in the SM so they would survive the journey back to Earth inside the CM.

After reentering the Earth's atmosphere, the three main parachutes opened up as usual; however, one of them collapsed before splashdown. Despite this, the CM had slowed enough and made a safe landing, albeit a little harder than usual, in the North Pacific on August 7, 1971, 295:11 GET.

APOLLO 16

This was the tenth manned mission to take flight in the Apollo program; its destination was the Descartes region of the Moon. The area is characterized by hilly, furrowed terrain, and was selected because it was thought to be representative of much of the Moon's surface. It was the first time that an area like this was explored.

Apollo 16 and its crew, Commander John Young, CM Pilot Ken Mattingly, and LM Pilot Charles Duke, launched on April 16, 1972, at 12:54 EST. After a safe journey, at 104:30 GET, the LM touched down on the Moon's surface about 276 m away from the landing site. The visit would provide Young and Duke with a seventy-one-hour stay on the lunar surface, during which they totalled more than twenty hours of moonwalks over the course of three EVAs. The priorities for their first EVA were to remove the LRV from stowage in the descent stage and activate its systems, as well as to deploy and set up another ALSEP.

The second and third EVAs were predominantly focused on geological exploration and sample-gathering in preselected areas. By the end of the mission, the LRV had travelled a total of 27.1 km, a little less than Apollo 15. In addition to the scientific experiments they were there to carry out, Young and Duke were the first to take photographs with an ultraviolet camera on the Moon. Since there was no atmospheric interference, they were able to capture images of celestial regions that are not visible from the Earth.

As with the previous mission, just before TEI they released a subsatellite into lunar orbit. This measured information on the Moon's mass, the interaction of the Moon's magnetic field and the Earth's, and the particle composition of space near the Moon. The three astronauts were then ready for the three-day journey home, during which Mattingly performed a spacewalk to retrieve the film cassettes from the SM. The CM would return to Earth and splash down in the South Pacific on April 27, 1972, 265:51 GET.

APOLLO 17

The final Apollo mission. Its crew members were the last humans to set foot on the Moon, and the last to travel beyond low-Earth orbit for nearly fifty years. Lift-off took place at 00:33 EST on December 7, 1972, the only night launch of the Apollo program. On board was Commander Eugene Cernan, accompanied by CM Pilot Ronald Evans and LM Pilot Harrison Schmitt.

Apollo 17's LM would be on the lunar surface longer than on any other mission, allowing Cernan and Schmitt to bring back the largest haul of samples yet, which meant that, in turn, Evans became the astronaut to have spent the longest consecutive amount of time in the Moon's orbit.

The LM and its crew touched down on the Moon just over 110 hours after the craft's launch from Earth. As with the two previous missions, there would be three EVAs, totalling more than twenty-two hours, during their three-day stay. Again the two astronauts would begin by deploying another ALSEP, before continuing on with more geology traverses. Apollo 17 was the only mission to bring with it the Traverse Gravimeter Experiment (TGE). Gravimeters are very sensitive, able to detect tiny changes in gravity. They had been highly effective in determining the geological makeup of Earth and helped us learn a great deal more about the Moon's internal composition. Cernan and Schmitt, who was a professional geologist, took twenty-six separate measurements with the TGE, which they carried aboard the LRV during the traverses. The LRV travelled farther than the two that had preceded it, clocking up 35.7 km in the four and a half hours it was operating.

Cernan and Schmitt lifted off from the surface at 22:54 UTC on December 14, and to this day remain the last people to set foot on the Moon. They travelled back to Evans in lunar orbit, and transferred themselves and their samples over to the CM. On the journey home it was Evans who carried out the spacewalk to retrieve the film from the SM, which took him one hour and seven minutes. The final time an Apollo CM reentered the Earth's atmosphere was on December 19, 1972, at 301:51 GET.

UNMANNED MISSIONS

NASA carried out many unmanned Apollo missions in order to test the functionality of the spacecraft and Saturn rockets. These missions would help minimize the danger to personnel by proving the equipment space-worthy. The first unmanned missions began in 1961, predominantly using the Saturn I to test the H1 engines and guidance systems. On its later outings the Saturn I carried with it a boilerplate CSM for analysis of the spacecraft's aerodynamics. Alongside these missions, NASA was also using a much smaller rocket, Little Joe II, to test the Launch Escape System and landing parachutes.

Prior to the disaster that became Apollo 1, there had been three test flights that incorporated a working CSM aboard a Saturn rocket, which is why NASA chose to resume the mission numbering with Apollo 4. The missions covered on these pages are the final test flights, which begin with the introduction of the Saturn 1B and operational CSMs.

AS 201

The first flight of the Block I Apollo CSM and the Saturn 1B launch vehicle. The Block I CSM differed from the Block II in that it was heavier, and was incapable of lunar landings because there was no docking hatch with which to dock with an LM. The first Block II CSM would be used on Apollo 7. This suborbital test flight demonstrated the capabilities of the Service Propulsion System, the Reaction Control Systems on both the CM and the SM, and the CM's ability to survive reentry.

AS 203

This test flight did not carry any Apollo CSM or LM, as its objective was to test the in-flight restart capability of the S-IVB rocket. For a lunar mission to succeed, the S-IVB had to shut down after placing the spacecraft in Earth's orbit, and it was then crucial that it would restart to fire the craft toward the Moon. In place of the CSM a streamlined nose cone had been fitted. The S-IVB sat atop the Saturn 1B as its second stage, but would ultimately become the third stage of the Saturn V.

AS 202

This was the second unmanned flight of the Block I CSM, and the third test flight of the Saturn 1B. This suborbital flight would further test the Saturn 1B's capabilities by launching the rocket higher, for a flight lasting twice as long as AS-201. It was the first flight to incorporate the Apollo spacecraft's Guidance and Navigation Control System. AS-202 took place after AS-203 because there had been a delay in preparing the Apollo spacecraft for the mission.

UNMANNED MISSION	LAUNCH VEHICLE	LAUNCH DATE
AS-201	Saturn IB	02.26.66
AS-203	Saturn IB	07.05.66
AS-202	Saturn IB	08.25.66
AS-204/Apollo 1	NA	01.27.67
Apollo 4	Saturn V	11.09.67
Apollo 5	Saturn IB	01.22.68
Apollo 6	Saturn V	04.04.68

APOLLO 4

Apollo 4 was the first test of the Saturn V launch vehicle, and the first operation of its first two stages (the S-IC and S-II). At the time of lift-off it was the largest launch vehicle to have ever flown. This flight was originally intended for launch in late 1966, but was delayed due to problems in the development of the S-II stage, and partially because of wiring faults NASA had discovered during the investigation in the wake of the Apollo 1 catastrophe.

APOLLO 5

This mission was the first to carry the LM into space. While in orbit the LM's ability to separate the descent and ascent stages was verified, as was each stage's propulsion systems. This mission was originally planned for April 1967, but similar to Apollo 4, there would be long delays due to technical issues. The launch vehicle used for this mission was the smaller Saturn 1B; although not as powerful as the Saturn V, it was adequate for placing the spacecraft in Earth orbit.

APOLLO 6

Apollo 6 was the final unmanned mission of the Apollo program. This mission established that the Saturn V launch vehicle and Apollo spacecraft were suitable for manned missions. Although there were some technical issues, the spacecraft was fired farther from Earth than on the previous missions, and the CM reentered Earth's atmosphere at just a little less than its target speed of 40,000 km/h. The CM was recovered 80 km from its planned destination in the North Pacific in good condition.

SKYLAB

Skylab was the United States' first space station. It performed as a solar observatory, an Earth-observing facility, and a microgravity lab. Other facilities included the workshop, kitchen, shower, toilet, sleeping bags, and exercise equipment, all specifically designed to function in microgravity. Skylab remained in orbit from 1973 until 1979.

In 1969, at roughly the same time as the first Moon landing, it was decided that Skylab would be launched into orbit using a Saturn V. Initial plans were drawn up for the space station to be built in orbit, carried up in pieces by several Saturn IB launches, but this idea was scrapped. A single Saturn V launch would be less costly, although it would be at the expense of a further Moon landing. Placing the space station in orbit would be the final launch of a Saturn V.

Skylab lifted off on May 14, 1973, from Pad 39A. As it tore through the atmosphere a meteoroid shield broke away, taking with it one of two solar panels and damaging the other. The incident deprived the space station of most of its electrical power, and threatened to render it useless. The top priority of the first crew to visit the station—who arrived aboard an Apollo CSM eleven days after Skylab made it to orbit—would be to make the necessary repairs and get the power back online. They conducted the first major repair in space over the course of two spacewalks, managing to restore power to the craft. The crew members spent twenty-eight days aboard Skylab, conducting medical experiments, gathering solar and Earth science data, and making further minor repairs. The crew's presence aboard the station set a new spaceflight duration record, a record that was then broken by each of the two subsequent Skylab missions. The next crew docked with Skylab on July 28, 1973. Problems with the Apollo SM thruster quads developing leaks became a huge concern for the astronauts' return journey. As Skylab had the ability to have two craft dock with it, NASA prepared for a rescue mission, and another Apollo spacecraft was readied at the launch complex. But it never launched, as it was ultimately deemed that the Apollo spacecraft in orbit would be able to make a safe return, which it did after fifty-nine days in space. The final crew to visit Skylab docked on November 16, 1973. The largest problem that they had to overcome was managing their workload. The crew felt strongly that they were being pushed too hard: Their daily instructions, sent by teleprinter, were sometimes fifteen meters long. After communicating their concerns to the ground crew, their workload was modified. Nevertheless, by the time they returned to Earth, they had completed even more work than had been planned before launch.

Skylab had originally been intended to remain in space much longer, but delays in preparing the Space Shuttle meant that Skylab's decaying orbit could not be stopped. Reentry began on July 11, 1979, and although NASA tried to direct its trajectory over the ocean, parts of Skylab rained down on Western Australia.

APOLLO–SOYUZ TEST PROJECT

The Apollo–Soyuz Test Project was the first joint United States–Soviet spaceflight. Tensions between the two countries had eased after the United States ended its involvement in the Vietnam War, and so a cooperative space mission was proposed as a symbol of their improving relationship. The mission would involve each nation sending a spacecraft into Earth orbit to dock with one another, allowing the crew to transfer between the two.

The two craft launched on July 15, 1975, with the Soviets launching first, followed by the Americans seven and a half hours later. Aboard Soyuz were two cosmonauts: Commander Alexey Leonov and Flight Engineer Valeri Kubasov, whereas the Apollo craft would, as usual, have three astronauts: Commander Thomas P. Stafford, Command Module Pilot Vance D. Brand, and Docking Module Pilot Deke Slayton. Over the course of the next two days, the Apollo and Soyuz made orbital adjustments until they were both brought into a circular orbit, at an altitude of 229 km. The Americans and Soviets docked together on July 17, with the hatches between the two craft being opened at 15:17 EST. The two mission commanders initially shook hands through the open hatches before the rest of the crews warmly greeted each other. President Gerald Ford and (an official acting on behalf of) Soviet Communist Party General Secretary Leonid Brezhnev both made congratulatory calls to the parties aboard the historic flight, and the crews exchanged gifts and had a meal together before closing the hatch for the day.

The next day was a busy one for the two crews. It began with a televised tour of each vehicle, before they commenced with a number of collaborative science experiments. Brand worked with Kubasov in the Soyuz, while Leonov joined Stafford and Slayton in the Apollo. After lunch Leonov and Stafford transferred to the Soyuz, swapping places with Brand and Kubasov. By late afternoon, once the crews had said their final speeches and good-byes (Stafford's drawl when speaking Russian was so pronounced that Leonov later joked that there were three languages spoken on the mission: Russian, English, and "Oklahomski"), they returned to their own ships and closed their hatches for the final time.

The Apollo and Soyuz spacecraft separated on July 19, after forty-four hours together. Moments after the undocking, the Apollo spacecraft intentionally shielded the Soyuz from the Sun. This allowed the Soviet cosmonauts to photograph the Sun's corona without receiving glare. Following this maneuver, the two spacecraft briefly docked once more, before separating again to go on their separate ways. Both would remain in orbit, conducting experiments and observing the Earth. Soyuz made a safe landing near its target of Baikonur Cosmodrome on July 21; Apollo splashed down a few days later on July 24, in the Pacific Ocean west of Hawaii.

MOONWALKERS

The men whom NASA selected to be astronauts had many things in common. They were all highly driven individuals, who throughout their lives pushed themselves hard to succeed. They all came from backgrounds in aviation, with many of them having flown dangerous Cold War missions, often carrying nuclear weapons behind enemy lines. Others had seen action in the skies during the Korean War, and many of them went on to become test pilots. Being a test pilot was seen as the pinnacle of a pilot's career: These were the people testing the latest, fastest, most maneuverable and high-tech equipment out there. The hair-raising experiences these men faced during their careers gave them the composure they needed to remain cool under pressure. All achieved Bachelor of Science degrees, and furthered them with master's degrees in Aeronautical or Astronautical Engineering; some even went on to get PhDs.

Aside from their obvious career-related similarities, there was another characteristic that these men shared: They were all obsessed with flying the highest, flying the farthest, but mostly, it seems, flying the fastest. Most of them got hooked on aviation from a young age, and the exhilaration they found in it developed into an addiction that lasted throughout their lifetimes. One notable exception was Harrison Schmitt, the only geologist to be assigned to an Apollo mission; science was the passion that consumed him.

The astronauts on the opposite page are the only twelve men to have walked on the Moon. They represent the best of the best, and while they shared many attributes, they also possessed individual strengths and skills that were necessary to form well-rounded, elite teams of space travellers.

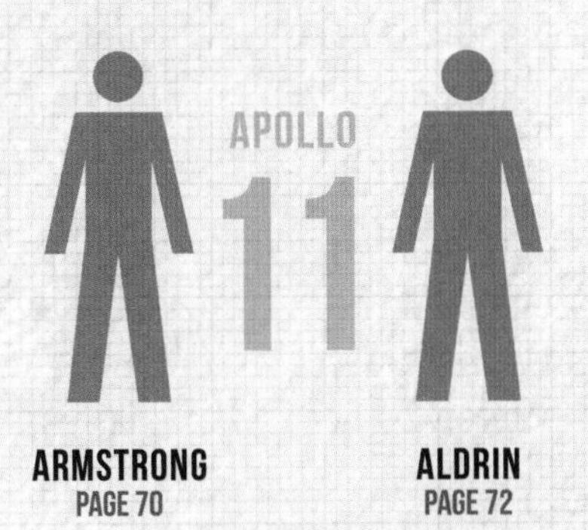
APOLLO
11
ARMSTRONG
PAGE 70
ALDRIN
PAGE 72

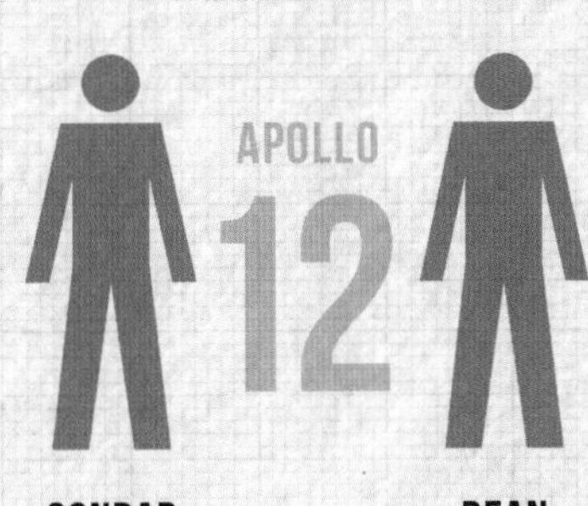
APOLLO
12
CONRAD
PAGE 74
BEAN
PAGE 76

APOLLO
14
SHEPARD
PAGE 78
MITCHELL
PAGE 80

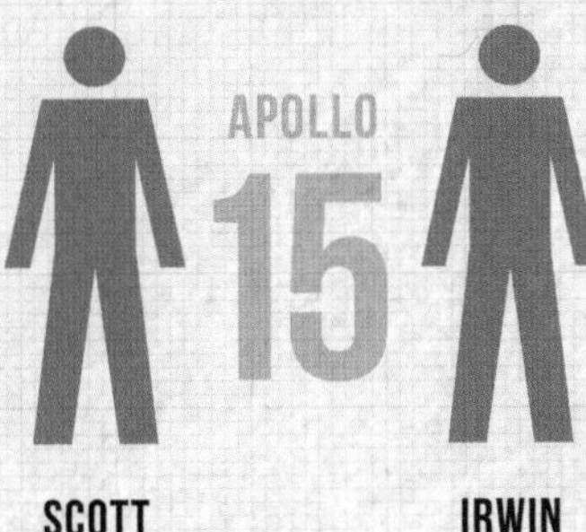
APOLLO
15
SCOTT
PAGE 82
IRWIN
PAGE 84

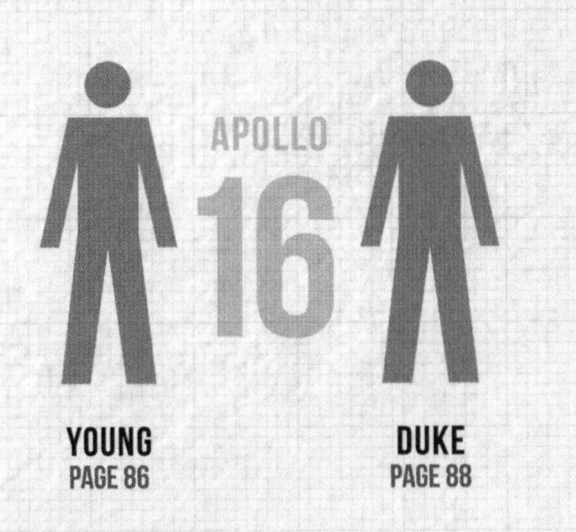
APOLLO
16
YOUNG
PAGE 86
DUKE
PAGE 88

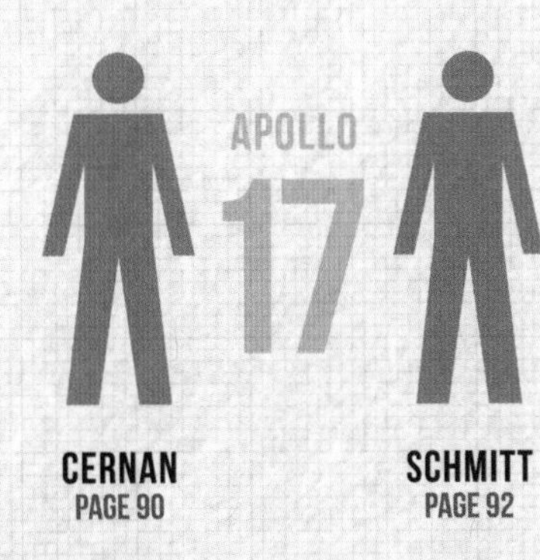
APOLLO
17
CERNAN
PAGE 90
SCHMITT
PAGE 92

NEIL ALDEN ARMSTRONG

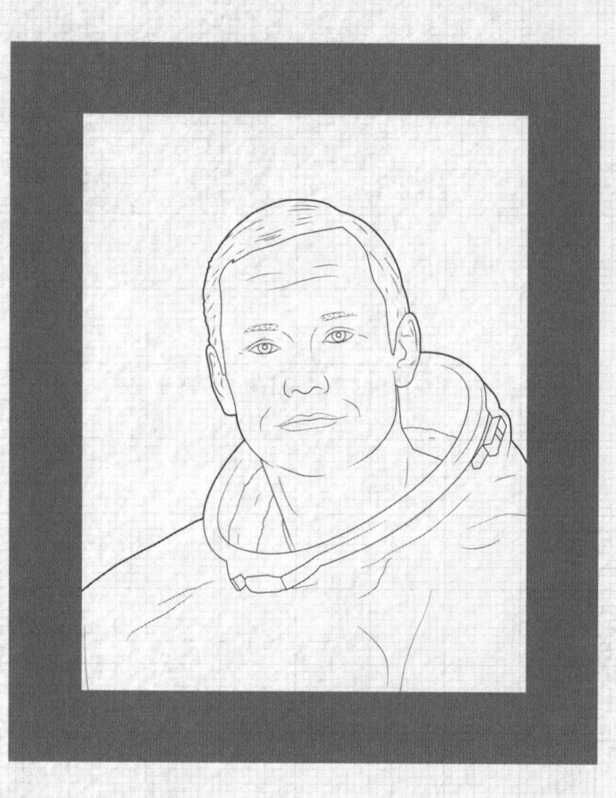

BORN:
AUGUST 5, 1930
WAPAKONETA, OHIO

Quiet, thoughtful, but utterly determined, Neil developed a keen interest in flying after he was taken to an air show when he was only two. When, a week shy of his sixth birthday, his father arranged for them to take a ride in a Ford Trimotor, he became completely hooked. While in high school, Neil took flying lessons and was an avid Boy Scout, where he gained the highest possible rank of Eagle Scout, an achievement that takes years to accomplish. On his sixteenth birthday, he attained a student flight certificate, and later that month he made his first solo flight.

In 1947, Neil enrolled at Purdue University to study Aeronautical Engineering. His studies were cut short when he was called up to serve in the Korean War as a pilot in the Navy.

PRESIDENTIAL MEDAL OF FREEDOM WITH DISTINCTION

CONGRESSIONAL SPACE MEDAL OF HONOR

NASA DISTINGUISHED SERVICE MEDAL

NATIONAL DEFENSE SERVICE MEDAL

KOREAN SERVICE MEDAL

ARMED FORCES RESERVE MEDAL

PRESIDENTIAL UNIT CITATION (KOREA)

UNITED NATIONS KOREA MEDAL

KOREAN WAR SERVICE MEDAL

After receiving eighteen months of flight training, he became a fully qualified Naval Aviator and, shortly after, left for Korea. He flew seventy-eight missions during his time there, was shot down once, and was awarded three medals.

On Neil's return from Korea in 1952, he left the Navy, though he remained in the US Naval Reserve until 1960. He returned to Purdue to complete his degree, graduating in 1955. He then took a job as an experimental research test pilot for the National Advisory Committee for Aeronautics (the precursor to NASA). Here he flew a huge array of aircraft, including the rocket-powered 6,000 km/h X-15. Over the course of his career, he flew in excess of two hundered models of aircraft.

A fast learner (a colleague remembered that Armstrong had a mind able to absorb knowledge like a sponge), Armstrong was selected for the Gemini space program. In 1965 he launched into orbit aboard Gemini 8, which accomplished the first docking of two spacecraft in orbit. His next and final role in the Gemini program was as the back-up Command Pilot for Gemini 11. Although not travelling into space on this mission, he was instrumental in the role of CAPCOM. Undoubtedly, his most well-known moment came on Apollo 11, landing a spacecraft on the Moon, and being the first person to set foot on another celestial body.

Not long after his return from the Moon, Neil announced that he did not intend to fly to space again, and he resigned from NASA in 1971. Famously humble, Neil has often been referred to as a "reluctant" American hero. He spent much of his later years teaching Aerospace Engineering and shunning the limelight.

SPACE MISSIONS:

Gemini 8
Apollo 11

TOTAL EVAs:

Totalling:
2h 31m

TIME IN SPACE:

Totalling:
8d 14h 12m

EDWIN EUGENE "BUZZ" ALDRIN JR.

BORN:
JANUARY 20, 1930
MONTCLAIR, NEW JERSEY

AIR FORCE DISTINGUISHED SERVICE MEDAL
LEGION OF MERIT
DISTINGUISHED FLYING CROSS
AIR MEDAL
AIR FORCE COMMENDATION MEDAL
OUTSTANDING UNIT AWARD
PRESIDENTIAL MEDAL OF FREEDOM WITH DISTINCTION
NASA DISTINGUISHED SERVICE MEDAL
NASA EXCEPTIONAL SERVICE MEDAL
NATIONAL DEFENSE SERVICE MEDAL
KOREAN SERVICE MEDAL
AIR FORCE LONGEVITY SERVICE AWARD
PRESIDENTIAL UNIT CITATION (KOREA)
UNITED NATIONS KOREA MEDAL
KOREAN WAR SERVICE MEDAL

Buzz got his nickname during childhood; one of his sisters would mispronounce brother as "buzzer," and this got shortened to Buzz—he made it his legal first name in 1988.

In high school Buzz had a real passion for football, but due to his father's influence he gave it up to focus on his studies. He graduated a year early and in 1947 went to study Mechanical Engineering at the United States Military Academy at West Point. He earned his Bachelor of Science degree in 1951, coming third in his class.

After West Point, Buzz was commissioned as an officer in the United States Air Force, initially serving as a fighter pilot in the Korean War. He flew sixty-six combat missions and shot down two enemy aircraft. On his return he was decorated with the Distinguished Flying Cross for his service during the war. Buzz continued in the Air Force, and by 1963 he had been awarded a Doctor of Science degree in Astronautics from MIT. In his graduate thesis he outlined plans for the space rendezvous and docking techniques that were critical to the success of the Apollo missions.

Buzz was selected to be an astronaut late in 1963, once it was no longer mandatory to have been a test pilot. He served as part of the back-up crew for Gemini 9A and was chosen to go on the final mission of the Gemini program, Gemini 12, in 1966. On this mission he performed more than five hours of EVAs while in Earth's orbit, setting new records for the time, while proving that astronauts could effectively work outside of their spacecraft. A few years later he was picked to be the Lunar Module Pilot on the monumental Apollo 11, cementing his place in history as part of the first lunar landing and one of the first two humans to walk on the Moon. Aldrin's first words on the Moon were "Beautiful view." When Armstrong then asked, "Isn't it magnificent?" Aldrin responded, "Magnificent desolation."

Buzz left NASA in 1971 and for a short period was assigned to the position of Commandant of the US Air Force Test Pilot School, before retiring from active duty after twenty-one years of service. Later in life Buzz became an enthusiastic ambassador for space exploration as well as a successful author, releasing works of science-fiction in addition to memoirs and children's books.

SPACE MISSIONS:

Gemini 12
Apollo 11

TOTAL EVAs:

Totalling:
7h 52m

TIME IN SPACE:

Totalling:
12d 1h 52m

CHARLES "PETE" CONRAD JR.

BORN:
JUNE 2, 1930
PHILADELPHIA, PENNSYLVANIA

NAVY DISTINGUISHED SERVICE MEDAL
DISTINGUISHED FLYING CROSS
CONGRESSIONAL SPACE MEDAL OF HONOR
NASA DISTINGUISHED SERVICE MEDAL
NASA EXCEPTIONAL SERVICE MEDAL
COLLIER TROPHY
HARMON TROPHY
THOMPSON TROPHY

It was clear from an early age that Pete was very bright, but he often struggled with his schoolwork. He had dyslexia, which was poorly understood at the time, and he was expelled when he failed most of his eleventh-grade exams. His mother found him a more suitable school, where he learned to work around being dyslexic, and he finally started to excel academically.

During his vactaions from school, Pete began working at Paoli Airfield doing odd jobs in return for flights and the occasional lesson. As he learned more about the workings of the aircraft, he progressed from doing odd jobs to carrying out maintenance work. Pete managed to earn his pilot's licence before completing high school. He ended up such a success at school that in 1949 he was not only accepted by the prestigious Princeton University, but was also awarded a full scholarship through the Navy. Pete studied for a Bachelor of Science degree in Aeronautical Engineering and graduated in 1953. Upon graduation he entered the Navy, performing outstandingly throughout naval flight school. Pete had an illustrious career in the Navy that saw him become an aircraft carrier pilot, flight instructor, and finally a test pilot.

In 1959, NASA invited Pete to take part in the selection process for the first group of astronauts for the Mercury program. All the candidates had to undergo what they deemed to be invasive, demeaning, and unnecessary medical and psychological testing. Pete was the only one to rebel, though, and "acted up" during psychological tests. He walked out of the selection process after dropping off his stool sample on the desk of the clinic's commanding officer. It was presented in a gift box with a red ribbon tied in a bow. NASA's notation on Pete's initial application reads "not suitable for long-duration flight."

Pete began training to be an astronaut in 1962, having been persuaded to reapply by his old friend Alan Shepard, who had flown on the Mercury missions. Pete found the testing much less intrusive this time and managed not to kick up a fuss. His first assignment was Gemini 5 in 1965, a mission that set a new endurance record of almost eight days in space. In 1966 he commanded Gemini 11, which carried out a successful docking procedure with an unmanned target vehicle in Earth orbit. Pete's big moment came on Apollo 12 in 1969, being part of the second Moon landing. His last space mission was to be Skylab 2 in 1973, where he became commander of the first crew to board the Skylab space station.

SPACE MISSIONS:

Gemini 5 & 11,
Apollo 12, Skylab 2

TOTAL EVAs:

Totalling:
12h 44m

TIME IN SPACE:

Totalling:
49d 3h 38m

ALAN LAVERN BEAN

BORN:
MARCH 15, 1932
WHEELER, TEXAS

NAVY ASTRONAUT WINGS

NAVY DISTINGUISHED SERVICE MEDAL

NASA DISTINGUISHED SERVICE MEDAL

REAR ADMIRAL WILLIAM S. PARSONS AWARD FOR SCIENTIFIC AND TECHNICAL PROGRESS

UNIVERSITY OF TEXAS DISTINGUISHED ALUMNUS AWARD AND DISTINGUISHED ENGINEERING GRADUATE AWARD

NATIONAL ACADEMY OF TELEVISION ARTS AND SCIENCES TRUSTEES AWARD

Alan grew up near an air base and as a result became fascinated with the aircraft he saw on a daily basis. His bedroom ceiling was adorned with model aircraft that were a testament to both his passion for flying and his creativity. Enthusiasm for flying lasted throughout Alan's school years and took him to the University of Texas, where he would study for a Bachelor of Science degree in Aeronautical Engineering. Alongside his studies Alan competed in gymnastics and diving, and he earned his degree in 1955.

Alan was commissioned into the Navy after completing university and commenced with flight training. Following training, he was assigned to a jet attack squadron in Jacksonville, Florida, where, at the age of twenty-four, he became the youngest member of squadron VA-44. On completing his four-year tour of duty with the squadron, he went on to attend the Navy Test Pilot School at Patuxent River, Maryland, where he would gain experience flying many types of naval aircraft. His instructor there was Pete Conrad, who was so impressed with Alan's talent that he was instrumental in getting him assigned to Apollo 12.

By 1962 Alan had left the Navy Test Pilot School and was based with another attack squadron in Florida, this time in Cecil Field. He decided that he wanted to be part of the exclusive new breed of test pilots, the astronauts, because, as he put it, "I thought it might be even more fun than flying airplanes." Alan applied to NASA that year, and although he made the final stages, he was unsuccessful. Unfazed, he determinedly applied again the following year.

NASA selected Alan to be an astronaut in 1963, and for his first assignment he was chosen to be part of the back-up crew for Gemini 10 in 1966. Alan then had to wait three more years to get his chance to go into space, which came on the Apollo 12 Moon landing. His next and final mission to take him to space was as commander of Skylab 3 in 1973, a record-setting mission that spent fifty-nine days in orbit and travelled more than 39 million kilometers. Alan's last major role for NASA was being appointed as back-up commander for the US crew on the Apollo—Soyuz mission in 1975.

Unlike those of his peers who ventured into politics or private industry after leaving NASA, when Bean resigned in 1981 he pursued a career as a painter. His art focuses on his experiences of spaceflight.

SPACE MISSIONS:

Apollo 12
Skylab 3

TOTAL EVAs:

Totalling:
10h 26m

TIME IN SPACE:

Totalling:
69d 15h 45m

ALAN BARTLETT "AL" SHEPARD JR.

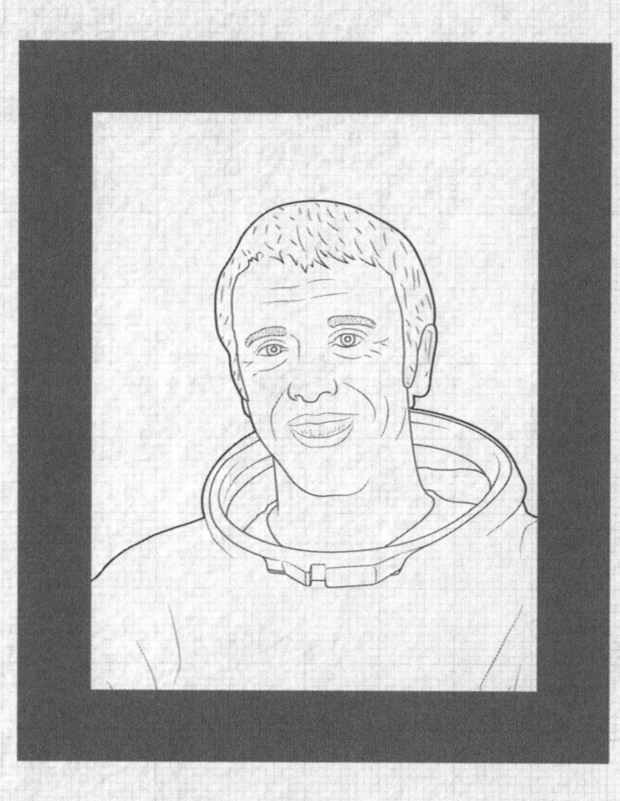

BORN:
NOVEMBER 18, 1923
DERRY,
NEW HAMPSHIRE

CONGRESSIONAL SPACE MEDAL OF HONOR
GOLDEN PLATE AWARD FOR SCIENCE AND EXPLORATION
LANGLEY GOLD MEDAL
JOHN J. MONTGOMERY AWARD

Alan was clever enough as a boy that he was able to skip both sixth and eighth grades. While he was in high school he developed an interest in flight; he was so enthused that he set up a model airplane club so he could share his passion with other students. In 1939 he started riding his bike to his local airfield so that he could do odd jobs in return for an occasional flight or informal lesson.

When the United States entered the Second World War, Alan's father encouraged him to join the US Army; however, Alan chose the Navy, effortlessly passing the entrance exam to the Naval Academy. Here he became a competitive sailor, rower, and swimmer, graduating in April 1944. Alan was posted to a destroyer in the Pacific, USS *Cogswell*, in

August 1944, where he experienced everything from kamikaze attacks to ship-sinking storms. When the war came to a close, Alan returned to the United States and began basic flight training at the start of 1946. Alan was awarded his Naval Aviator wings in 1947 and assigned to Fighter Squadron 42. He was attached to this squadron for a few years before being selected to attend the US Naval Test Pilot School, from which he graduated in 1951. As a qualified test pilot, he was experimenting with new in-flight refuelling techniques and, among other things, high-altitude aircraft.

In 1959 NASA chose the hard-living Alan as one of its first astronauts and subsequently for the Mercury 3 mission, which would carry the first American into space. Alan took his new role so seriously that he quit smoking and started jogging, though he was unable to renounce either cocktails or philandering. On May 5, 1961, just twenty-three days after Yuri Gagarin made the first human spaceflight (on learning of the news Alan slammed a table so hard that a watching colleague feared he may have broken his fist), Alan succeeded in his, a fifteen-minute flight that reached an altitude of 187km. Due to a problem that developed in his inner ear, Alan's next mission into space came nearly a decade later. His problem was eventually corrected using a new surgical technique, and Alan was assigned as the commander of the 1971 Apollo 14 mission. At forty-seven, he was the oldest man to set foot on the Moon, and he remains the only person to have played golf there.

In his later years Alan made a fortune in banking and real estate, and served on the boards of several corporations. He also cofounded and chaired the Mercury 7 Foundation, which offers college scholarships to those interested in science and engineering.

SPACE MISSIONS:

Mercury-Redstone 3
Apollo 14

TOTAL EVAs:

Totalling:
9h 23m

TIME IN SPACE:

Totalling:
9d 0h 57m

EDGAR DEAN "ED" MITCHELL

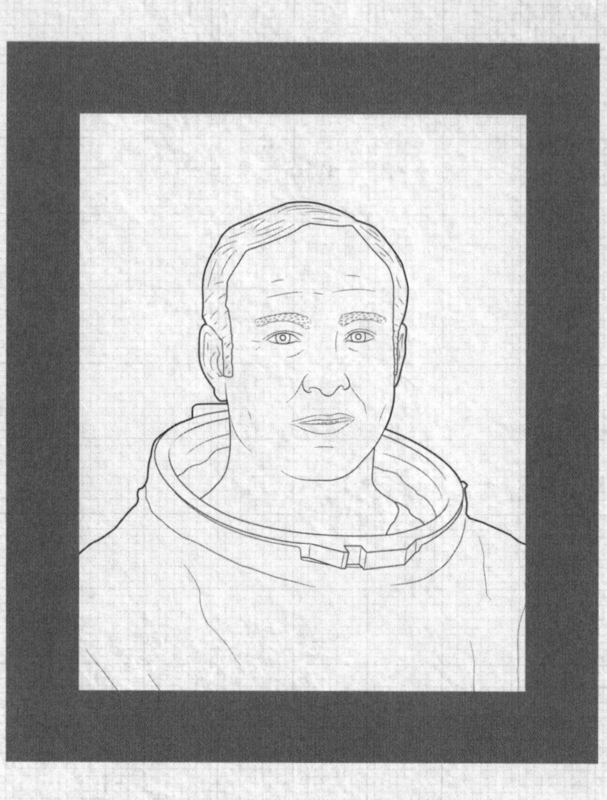

BORN:
SEPTEMBER 17, 1930
HEREFORD, TEXAS

PRESIDENTIAL MEDAL OF FREEDOM

MANNED SPACECRAFT CENTER SUPERIOR ACHIEVEMENT AWARD

ARNOLD AIR SOCIETY'S JOHN F. KENNEDY AWARD

NAVY ASTRONAUT WINGS

NAVY DISTINGUISHED SERVICE MEDAL

NASA DISTINGUISHED SERVICE MEDAL

Edgar grew up on a cattle ranch not far from Roswell, New Mexico. He took his first flight at the age of four when a barnstorming pilot brought his aircraft down in a nearby field and asked his parents for fuel. In return, the young Ed was treated to a ride in the plane, instilling in him a love of flight that would last his lifetime.

Ed attended the Carnegie Institute of Technology and graduated with a Bachelor of Science degree in Industrial Management in 1952. He entered the Navy later that year, completing his basic training at the San Diego Recruit Depot, then qualifying as a Naval Aviator in 1954. While serving in the Navy, Ed gained a wealth of experience piloting all sorts of aircraft, from land-based patrol planes and strategic bombers to carrier-based jet aircraft. He then qualified as a research pilot and was assigned to Air Development Squadron Five.

Ed recalls the moment he set his goals on becoming an astronaut: "I made the decision in 1957 when Sputnik went up." Knowing humans would be next into space, he began directing all his efforts toward this ambition. Ed studied for a bachelor's degree in Aeronautics at the US Naval Postgraduate School, which he attained in 1961, before going on to earn a doctorate in Aeronautics and Astronautics from MIT in 1964. He then attended the Aerospace Research Pilot School, qualifying as a test pilot in 1966, coming out top of his class. During this time Ed was also acting as an instructor in advanced mathematics and navigation theory for astronaut candidates.

Shortly after he became a certified test pilot, NASA selected Ed to be an astronaut. He was initially chosen as a member of the support crew for Apollo 9, and then designated as the back-up Lunar Module pilot on Apollo 10. Ed played a very important role as part of the Mission Operations Team for Apollo 13, working out strategies in the simulator to bring the ailing craft home. For his efforts he was awarded the Presidential Medal of Freedom by President Nixon in 1970. In 1971 he finally got to realize his dream when he was designated as the Lunar Module pilot on Apollo 14. His trip to the Moon would be his one and only venture into space. During the mission he experienced a "flash of understanding" that switched him on to the universe, sensing an intelligence he would spend the rest of his life trying to understand.

SPACE MISSIONS:

Apollo 14

TOTAL EVAs:

Totalling:
9h 23m

TIME IN SPACE:

Totalling:
9d 0h 01m

DAVID RANDOLPH “DAVE” SCOTT

BORN:
JUNE 6, 1932
SAN ANTONIO, TEXAS

NASA DISTINGUISHED SERVICE MEDAL
NASA EXCEPTIONAL SERVICE MEDAL
AIR FORCE DISTINGUISHED SERVICE MEDAL
DISTINGUISHED FLYING CROSS
AIR FORCE ASSOCIATION'S DAVID C. SCHILLING TROPHY
FÉDÉRATION AÉRONAUTIQUE INTERNATIONALE GOLD MEDAL
UNITED NATIONS PEACE MEDAL
ROBERT J. COLLIER TROPHY

Dave was born at Randolph Field (the source of his middle name), the Air Force base at which his father served. He became hooked on flight when, at three years old, he saw his father formation-flying in a biplane.

Dave attended the Western High School in Washington, DC, where he was a devoted member of the swimming team and set a handful of state records. His father pushed him to succeed, and Dave's efforts paid off when he was granted a scholarship to West Point Military Academy. He graduated in 1954 with a Bachelor of Science degree, and distinguished himself by coming fifth out of 633 students. Due to his excellent performance, he was given the choice of which branch of the military to serve. Desperate to fly jets, he joined the Air Force. Dave was initially assigned to Webb Air Force Base, Texas, where he completed his pilot training in 1955. He then proceeded with further gunnery training out of Laughlin Air Force Base, Texas, followed by Luke Air Force Base, Arizona. Dave's first tour of duty took him to Soesterberg Air Base in the Netherlands, where he was based from 1956 until 1960. His role here involved flying supersonic jets in Cold War missions over Europe.

On completing his tour, Dave returned to the United States to study. He attended MIT and by 1962 had attained two more qualifications: a Master of Science degree in Aeronautics/Astronautics and an engineering degree in Aeronautics/Astronautics. Dave then went on to train as a test pilot, graduating from the Air Force Experimental Test Pilot School in 1963 and the Aerospace Research Pilot School in 1964.

NASA selected Dave in 1963, and in 1966 he was chosen for Gemini 8 with Neil Armstrong. Although the mission was a success in other respects, Dave didn't get to carry out his planned spacewalk due to a defective thruster. In 1969 he was assigned as the Command Module pilot of Apollo 9, the first launch of a fully configured Apollo spacecraft. Dave's third spaceflight came in 1971, as commander of Apollo 15, when he not only got to walk on the Moon, but also be part of the first crew to use the Lunar Roving Vehicle.

SPACE MISSIONS:

Gemini 8
Apollo 9 & 15

TOTAL EVAs:

Totalling:
18h 35m

TIME IN SPACE:

Totalling:
22d 18h 53m

JAMES BENSON "JIM" IRWIN

BORN:
MARCH 17, 1930
PITTSBURGH, PENNSYLVANIA

COMMAND PILOT ASTRONAUT WINGS
AIR FORCE DISTINGUISHED SERVICE MEDAL
AIR FORCE COMMENDATION MEDAL
NASA DISTINGUISHED SERVICE MEDAL
UNITED NATIONS PEACE MEDAL
AIR FORCE ASSOCIATION'S DAVID C. SCHILLING TROPHY
ROBERT J. COLLIER TROPHY
HALEY ASTRONAUTICS AWARD
ARNOLD AIR SOCIETY'S JOHN F. KENNEDY TROPHY

Jim's interest in flying began around the age of six, when a neighbor gave him a model airplane that he cherished. His enthusiasm escalated throughout his youth as his father, who was a plumber by trade, took him on outings to a local airfield to watch the planes take off and land. Jim attended high school in Salt Lake City, Utah, and graduated in 1947.

Jim entered the United States Naval Academy in 1951, earning his Bachelor of Science degree in Naval Science. This was immediately followed by his flight training in Hondo, Texas. Jim learned so quickly that he soon found himself completely unchallenged by the T-6 training planes, and by the time he'd officially become a Naval Aviator, he was even thinking about leaving the service to work for an airline. This all changed, however, when he encountered his first P-51 fighter plane. The sheer exhilaration he got from piloting it gave him new drive, and from then on he found himself "living to fly."

Seeking new thrills, Jim decided to become a test pilot. To do this he needed a master's degree, and so he attended the University of Michigan to study Aeronautical Engineering and Instrumentation Engineering. Jim attained his qualification in 1957 and began training in 1960 at the Air Force Experimental Test Pilot School in California. Shortly after being certified he was assigned as a developmental test pilot for the Lockheed YF-12, which was top-secret at the time, as it was the highest-flying, fastest plane ever built. Jim's future was put on hold, however, when, in 1961, a student he was teaching crashed the plane they were in. They both survived, but Jim suffered amnesia and compound fractures, and nearly lost a leg. He made a full recovery and resumed flying in 1962. The following year he completed training at the Aerospace Research Pilot School, which enabled him to test an even greater array of cutting-edge equipment.

Jim was selected to be an astronaut by NASA in 1966. His first role came on Apollo 10, as a member of the support crew, and on Apollo 12 he was the designated back-up Lunar Module pilot. His only mission into space came in 1971, aboard Apollo 15.

SPACE MISSIONS:

Apollo 15

TOTAL EVAs:

Totalling:
18h 35m

TIME IN SPACE:

Totalling:
12d 7h 12m

JOHN WATTS YOUNG

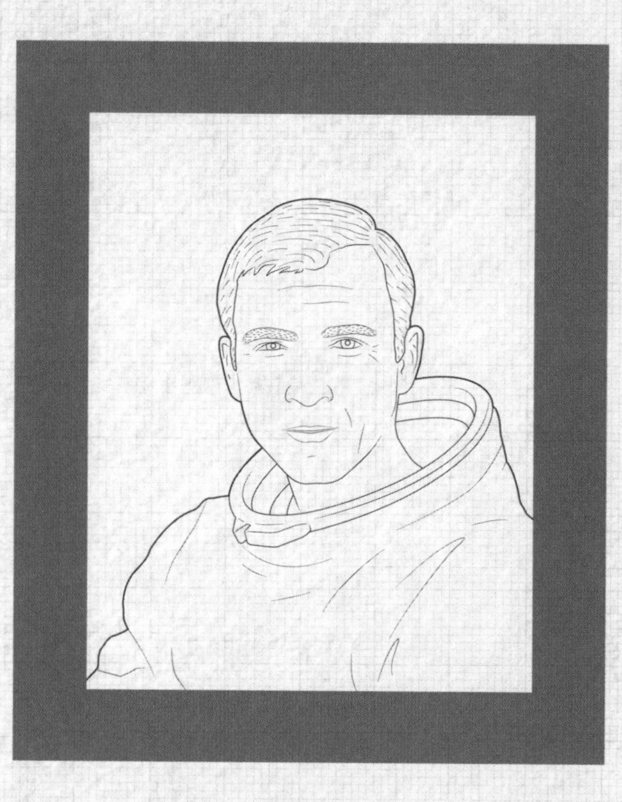

BORN:
SEPTEMBER 24, 1930
SAN FRANCISCO, CALIFORNIA

GENERAL JAMES E. HILL LIFETIME SPACE ACHIEVEMENT AWARD

GOLDEN PLATE AWARD FOR SCIENCE AND EXPLORATION

AMERICAN ASTRONAUTICAL SOCIETY SPACE FLIGHT AWARD

NASA AMBASSADOR OF EXPLORATION

NASA DISTINGUISHED SERVICE MEDAL

The Great Depression forced John and his family to leave California when he was just eighteen months old. They moved on to Georgia and then Florida, where they settled and John attended high school. Here he combined his twin passions of reading and building model aircraft. John graduated from high school in 1948 and went on to study for a Bachelor of Science degree in Aeronautical Engineering from the Georgia Institute of Technology. He graduated with highest honors in 1952.

Following university John joined the Navy, initially serving in Korea before receiving his flight training. For a short period he was designated as a helicopter pilot, before spending the next four years flying a variety of fighter planes from USS *Forrestal*.

Once certified as a test pilot in 1959, he was assigned to the Naval Air Test Center, where, over the next three years, he got to experiment with the latest military hardware. During his time here he set two time-to-climb world records, hitting altitudes of 3,000 m and 25,000 m from a runway start faster than anyone in history. He was such a gifted pilot that one of his contemporaries observed that he flew as if he wore the plane.

John joined NASA as an astronaut in 1962 and was assigned his first space mission in 1965: Gemini 3, the first manned launch of the Gemini program. John, landing another first, smuggled a corned-beef sandwich on board. Mid-mission, while they were supposed to be evaluating the space food, John took the sandwich out—much to the amazement (and amusement) of Virgil Grissom, his copilot. Even though he put it away after only one bite (bits of it had begun floating around in microgravity), he was heavily reprimanded for the stunt upon their return.

After the sandwich incident, John's future looked uncertain for a while. Nevertheless, after acting as back-up pilot on Gemini 6A, he got to travel into space once more, as commander of Gemini 10 in 1966. Later that year he was called up for the Apollo program. He was the designated back-up Command Module pilot on Apollo 7, and then the actual Command Module Pilot on Apollo 10, the landing "dress rehearsal," when he became the first man to orbit the Moon alone. He was involved with Apollo 13 as the back-up commander, but was not granted the privilege of exploring the Moon's surface until Apollo 16. After Apollo, John was part of the Space Shuttle program and commanded its maiden flight, STS-1, in 1981. His final journey into space came in 1983, on STS-9. John has had the longest career of any astronaut, having worked for NASA for forty-two years.

SPACE MISSIONS:

Gemini 3 & 10,
Apollo 10 & 16,
STS-1, STS-9

TOTAL EVAs:

Totalling:
20h 14m

TIME IN SPACE:

Totalling:
34d 19h 35m

CHARLES MOSS "CHARLIE" DUKE JR.

BORN:
OCTOBER 3, 1935
CHARLOTTE, NORTH CAROLINA

NASA DISTINGUISHED SERVICE MEDAL
JSC CERTIFICATE OF COMMENDATION
AIR FORCE DISTINGUISHED SERVICE MEDAL
LEGION OF MERIT
AIR FORCE COMMAND PILOT ASTRONAUT WINGS
SOCIETY OF EXPERIMENTAL TEST PILOTS' IVEN C. KINCHELOE AWARD
AMERICAN ASTRONAUTICAL SOCIETY FLIGHT ACHIEVEMENT AWARD
AMERICAN INSTITUTE OF AERONAUTICS AND ASTRONAUTICS HALEY ASTRONAUTICS AWARD

Charlie went to high school in Lancaster, South Carolina, before transferring to Admiral Farragut Academy in Florida, where he became the highest-achieving student in his year. Led by a desire to serve his country, the tall, handsome Duke joined the United States Naval Academy after graduation and completed his Bachelor of Science degree in Naval Sciences in 1957.

Charlie's initial flight training after being commissioned into the Air Force was carried out at Spence Air Base, Georgia, and concluded at Webb Air Force Base, Texas. He graduated with distinction from flight training in 1958, then continued with advanced flight training, where he was also a distinguished graduate. Charlie's first assignment in the Air Force was as a fighter pilot with the 526th Fighter-Interceptor Squadron in Germany, where he remained for three years.

On his return from Germany, Charlie went to the Aerospace Research Pilot School and qualified in 1965. He stayed on to become an instructor, teaching control systems, and gained more experience flying an assortment of fast jets.

In 1966 Charlie was chosen to be an astronaut on the Apollo program. He was part of the support crew for Apollo 10 and was assigned the role of CAPCOM for the first lunar landing, Apollo 11. His distinctive southern drawl was heard all over the world as he responded to Armstrong's first words on touching down: "Roger, Twank . . . Tranquillity, we copy you on the ground. You got a bunch of guys about to turn blue. We're breathing again. Thanks a lot." His next Apollo assignment would be as back-up Lunar Module pilot for Apollo 13. During training for this mission, Charlie caught the measles and unknowingly exposed both the prime and back-up crews to it. Ken Mattingly, on the prime crew, was the only one with no immunity to the disease and was replaced by Jack Swigert.

Charlie's moment came when he was assigned to Apollo 16, where he got to spend three days on the lunar surface on the second of the extended missions. Charlie would also be involved in the final mission, Apollo 17, as back-up Lunar Module pilot.

Charlie retired in 1975 from the astronaut program to conduct his private business, though he is currently Chairman of the Board of Directors of the Astronaut Scholarship Foundation. He became a committed Christian, with a particular interest in prison ministry.

SPACE MISSIONS:

Apollo 16

TOTAL EVAs:

Totalling:
20h 15m

TIME IN SPACE:

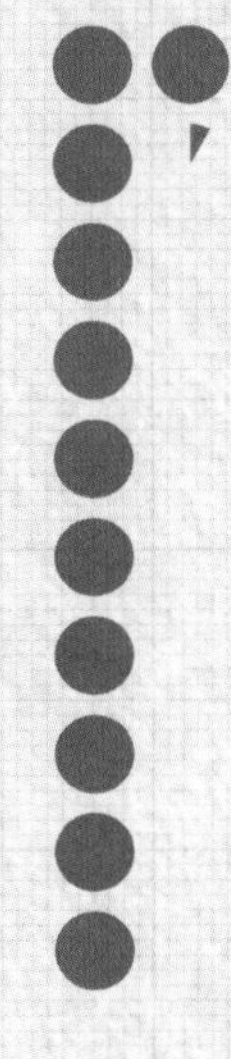

Totalling:
11d 1h 51m

EUGENE ANDREW "GENE" CERNAN

BORN:
MARCH 14, 1934
BELLWOOD, ILLINOIS

NAVAL AVIATOR ASTRONAUT INSIGNIA
NAVY DISTINGUISHED SERVICE MEDAL
DISTINGUISHED FLYING CROSS
NATIONAL DEFENSE SERVICE MEDAL
NASA DISTINGUISHED SERVICE MEDAL
NASA EXCEPTIONAL SERVICE MEDAL

When Gene saw footage of planes taking off from aircraft carriers, flying immediately became his childhood dream. Gene went on to attend Purdue University, with the hope of not only getting a degree but also of being commissioned into the Navy. He received his Bachelor of Science degree in Electrical Engineering in 1956 and entered the Navy the same year.

Gene qualified as a Naval Aviator in 1957 and went on to fly fighter jets with Attack Squadrons 126 and 113 at Miramar, California. With the escalation of the Space Race, the prospect of flying a rocket ship into space instantly appealed to Gene. He had not yet applied to be an astronaut when one day, out of the blue, he received a telephone call from a high-ranking Navy officer asking if he would like to be considered for the Apollo program. It took a moment for the enormity of the offer to sink in, before he hurriedly and enthusiastically accepted.

Gene's first spaceflight came in 1966 aboard Gemini 9A. Initially selected as back-up for this mission, he had to step up when the two prime crew were killed in a plane crash. The mission encountered several technical problems, but the crew was still able to simulate some of the procedures used on the Apollo missions.

In 1969 Gene was assigned as the Lunar Module pilot on Apollo 10, the "dress rehearsal" for the first Moon landing, where he performed the third ever spacewalk. His ultimate journey into space would be on Apollo 17, also the concluding mission of the program. Gene remains the last person to have set foot on the Moon, and, as he ascended the ladder to the LM, he also uttered the final human words ever spoken on the Moon. "Bob, this is Gene, and I'm on the surface; and, as I take man's last step from the surface, back home for some time to come—but we believe not too long into the future—I'd like to just say what I believe history will record: that America's challenge of today has forged man's destiny of tomorrow. And, as we leave the Moon at Taurus-Littrow, we leave as we came and, God willing, as we shall return, with peace and hope for all mankind. Godspeed the crew of Apollo 17."

Gene retired from NASA and the Navy in 1976, going into private business. He did, however, remain involved in space as a motivational speaker, and often provided commentary for television broadcasts of the Space Shuttle.

SPACE MISSIONS:

Gemini 9A
Apollo 10 & 17

TOTAL EVAs:

Totalling:
24h 11m

TIME IN SPACE:

Totalling:
23d 14h 15m

HARRISON HAGAN "JACK" SCHMITT

BORN:
JULY 3, 1935
SANTA RITA,
NEW MEXICO

NASA DISTINGUISHED SERVICE MEDAL
G. K. GILBERT AWARD
LEIF ERIKSON EXPLORATION AWARD

Jack was the first (and only) person to walk on the Moon who hadn't come from a background in aviation. As a geologist he possessed skills that would benefit the Apollo program in ways the pilots weren't able to.

Jack grew up in a small town in New Mexico. Like the other astronauts, he showed a lot of promise as a child, but unlike other astronauts it wasn't aircraft, jets, or rockets that spurred him on. His father, a mining geologist, introduced him to the sciences at a young age, and Jack used to assist him on weekends doing fieldwork. Jack studied geology at the California Institute of Technology and received his Bachelor of Science degree in 1957. He then travelled to the University of Oslo to carry out field studies, the results of which would earn him a PhD from Harvard in 1964.

Following his studies Jack returned to the United States to work for the United States Geological Survey in New Mexico and Montana, before landing a job at their Astrogeology Center in Flagstaff, Arizona. Here, Jack and other geologists went on field trips with the astronauts to teach them about geological formations and what to look out for on the Moon. During this time Jack was also working on geological field techniques that would be used on the Apollo missions. When Jack heard that NASA was looking for scientist-astronauts, he didn't hesitate in applying. He was successful but, with no background as a pilot, had to undergo an intensive fifty-three-week flight training course. Jack was certified by the Air Force as a jet pilot in 1965, and later, in 1967, the US Navy certified him as a helicopter pilot. He was still involved with the Apollo program during his flight training, helping to select landing sites and developing the field equipment to use once they got there.

Jack was originally intended for Apollo 18, but because of the cutbacks and cancellation of Apollo 18 onward, he was reassigned to Apollo 17. It was deemed important that, before the program came to a close, a professional geologist should be able to study the Moon firsthand. He collected what is widely regarded as the most interesting sample of Moon rock of the program, giving us greater insight into lunar history, including the likelihood that the Moon once possessed an active magnetic field. Jack retired from NASA in 1975 to serve as a Republican senator and has subsequently worked as a consultant in business, geology, space, and public policy.

SPACE MISSIONS:

Apollo 17

TOTAL EVAs:

Totalling:
22h 4m

TIME IN SPACE:

Totalling:
12d 13h 52m

MORE

Over the course of
the Apollo missions:

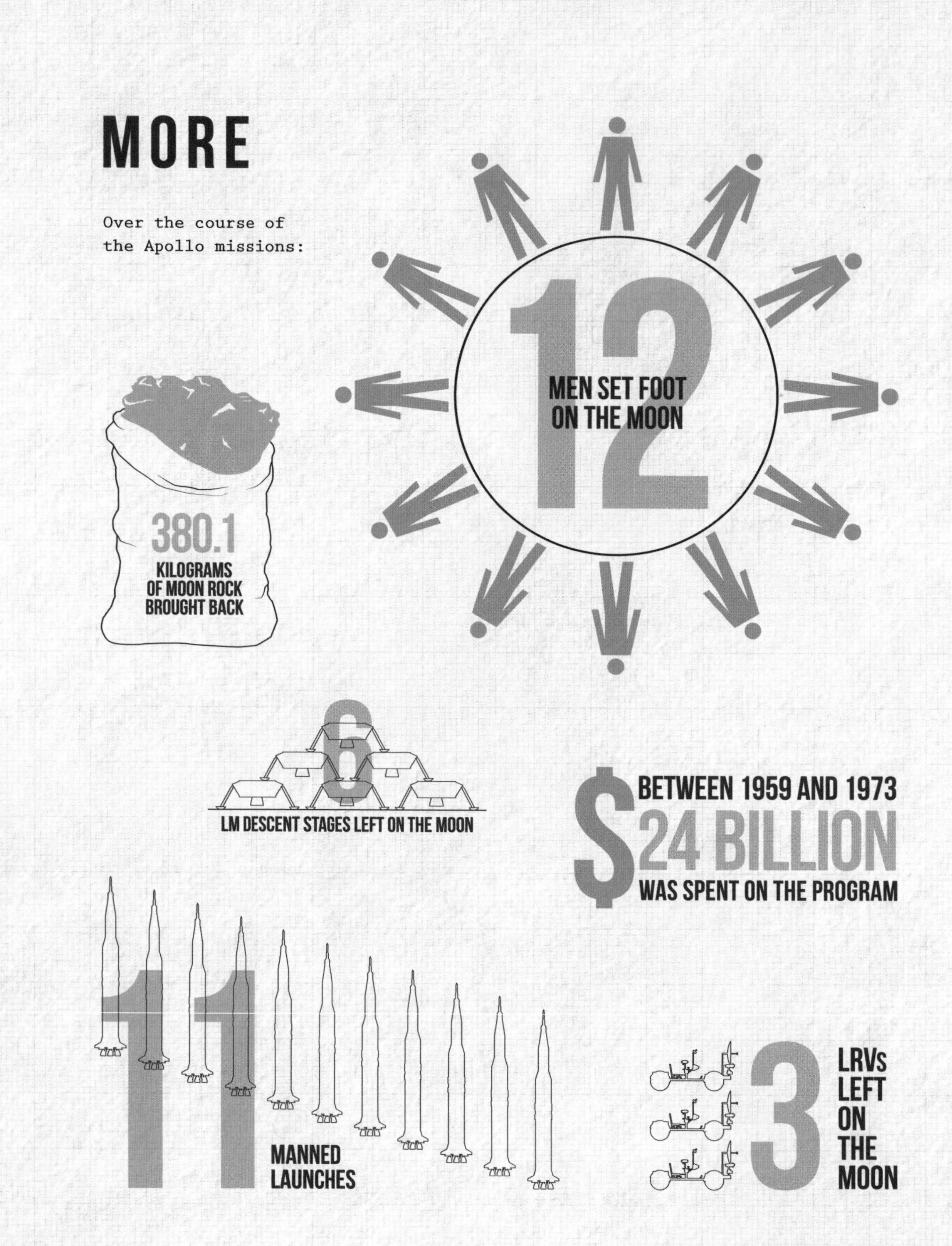

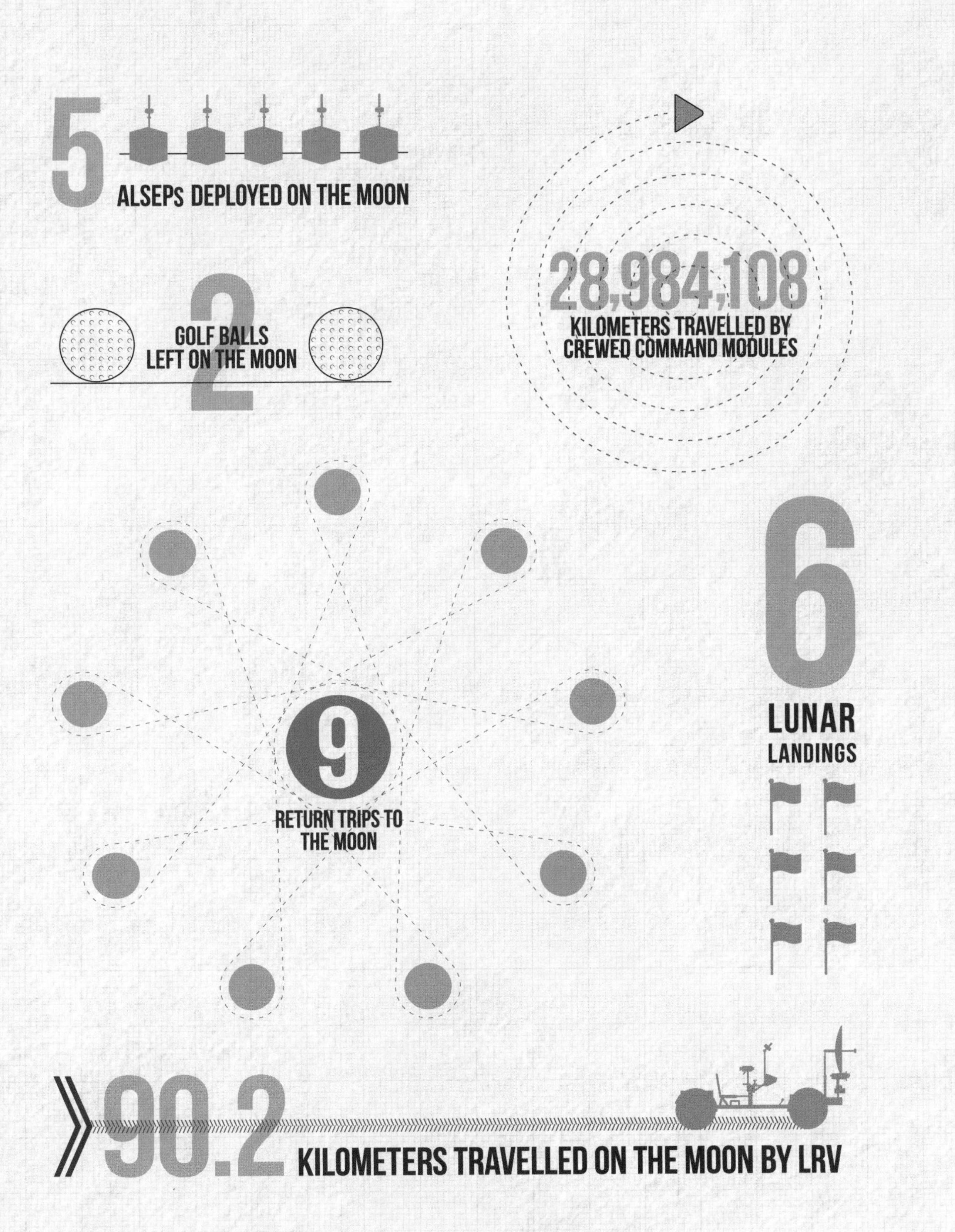
5
ALSEPs DEPLOYED ON THE MOON
28,984,108
KILOMETERS TRAVELLED BY CREWED COMMAND MODULES
2
GOLF BALLS LEFT ON THE MOON
6
LUNAR LANDINGS
9
RETURN TRIPS TO THE MOON
90.2
KILOMETERS TRAVELLED ON THE MOON BY LRV

MASS COMPARISONS

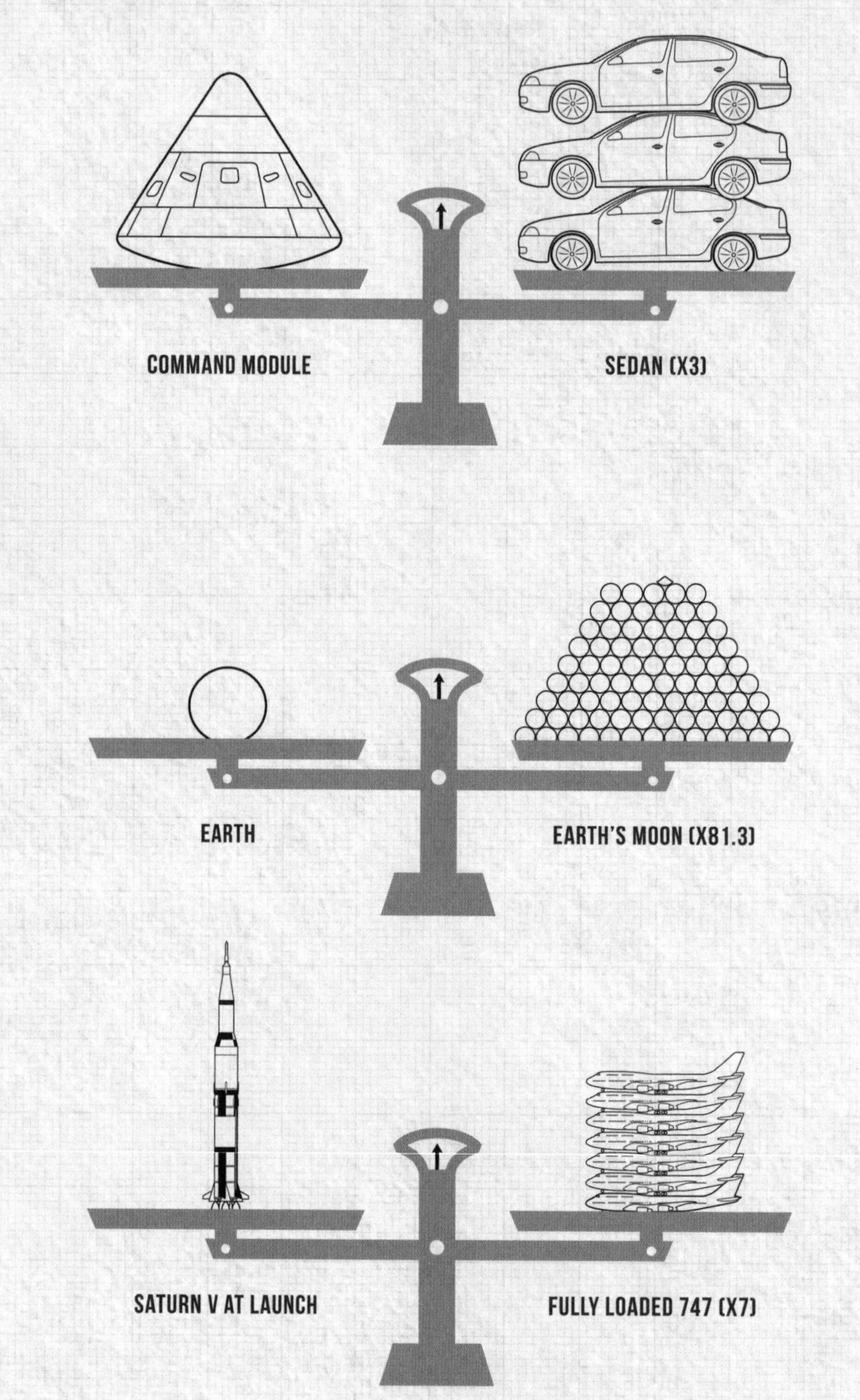

MOON TREES

On Apollo 14, Stuart Roosa took five hundred seeds from different species of tree to the Moon. They would be germinated on their return and planted throughout the United States of America and the rest of the world.

REFLECTED LIGHT

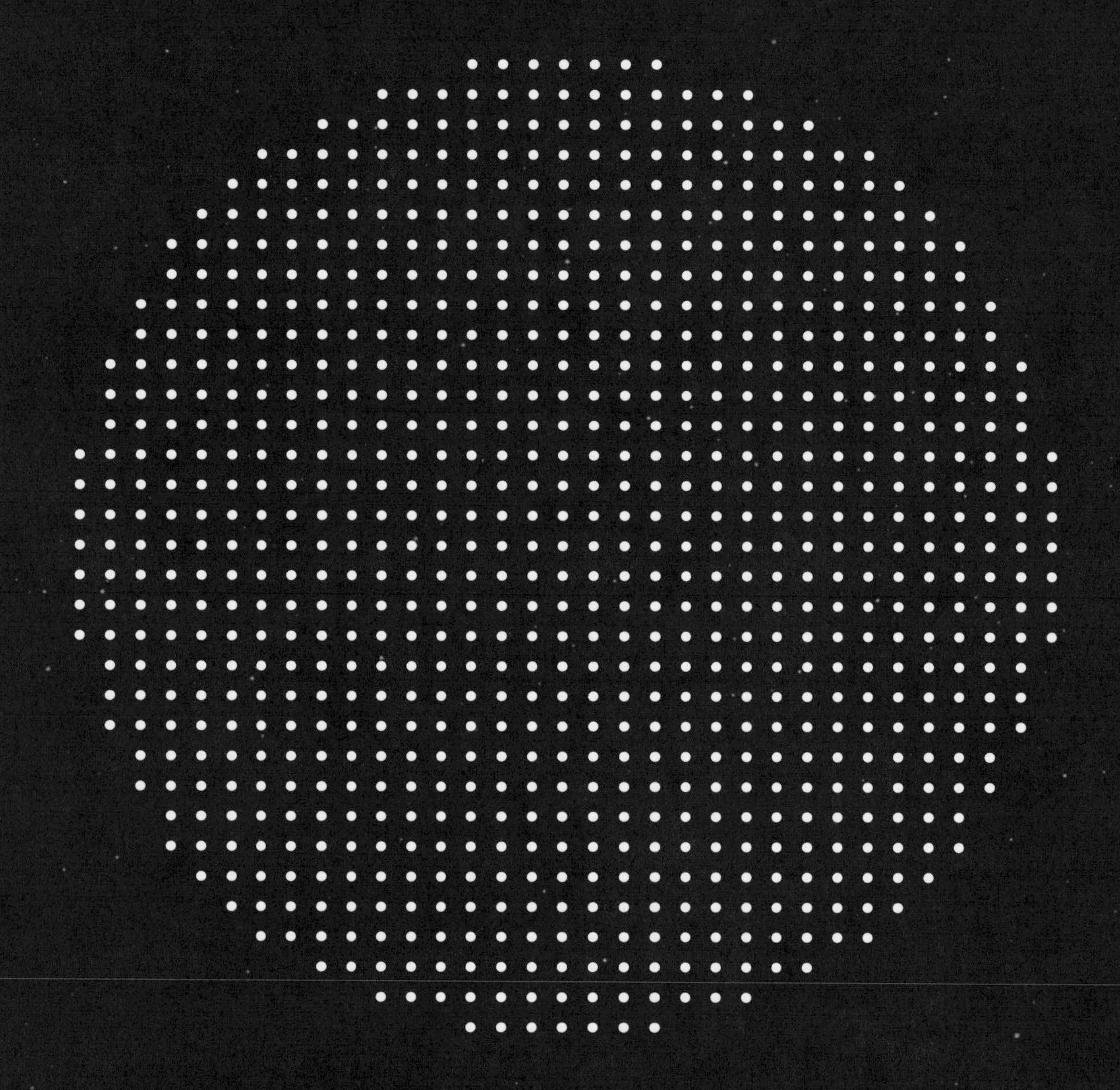

The Moon reflects
7 percent of visible light.

THE MOON'S ORBIT

The Moon orbits Earth elliptically, and therefore its distance from Earth varies. The perigee is the point at which the Moon is closest to Earth, and the apogee is when it's at its farthest.

SUPERMOONS AND MICROMOONS

At its perigee the Moon appears 14 percent bigger from Earth and 30 percent brighter. This is known as a Supermoon.

At its apogee the Moon appears smaller and dimmer, and is known as a Micromoon.

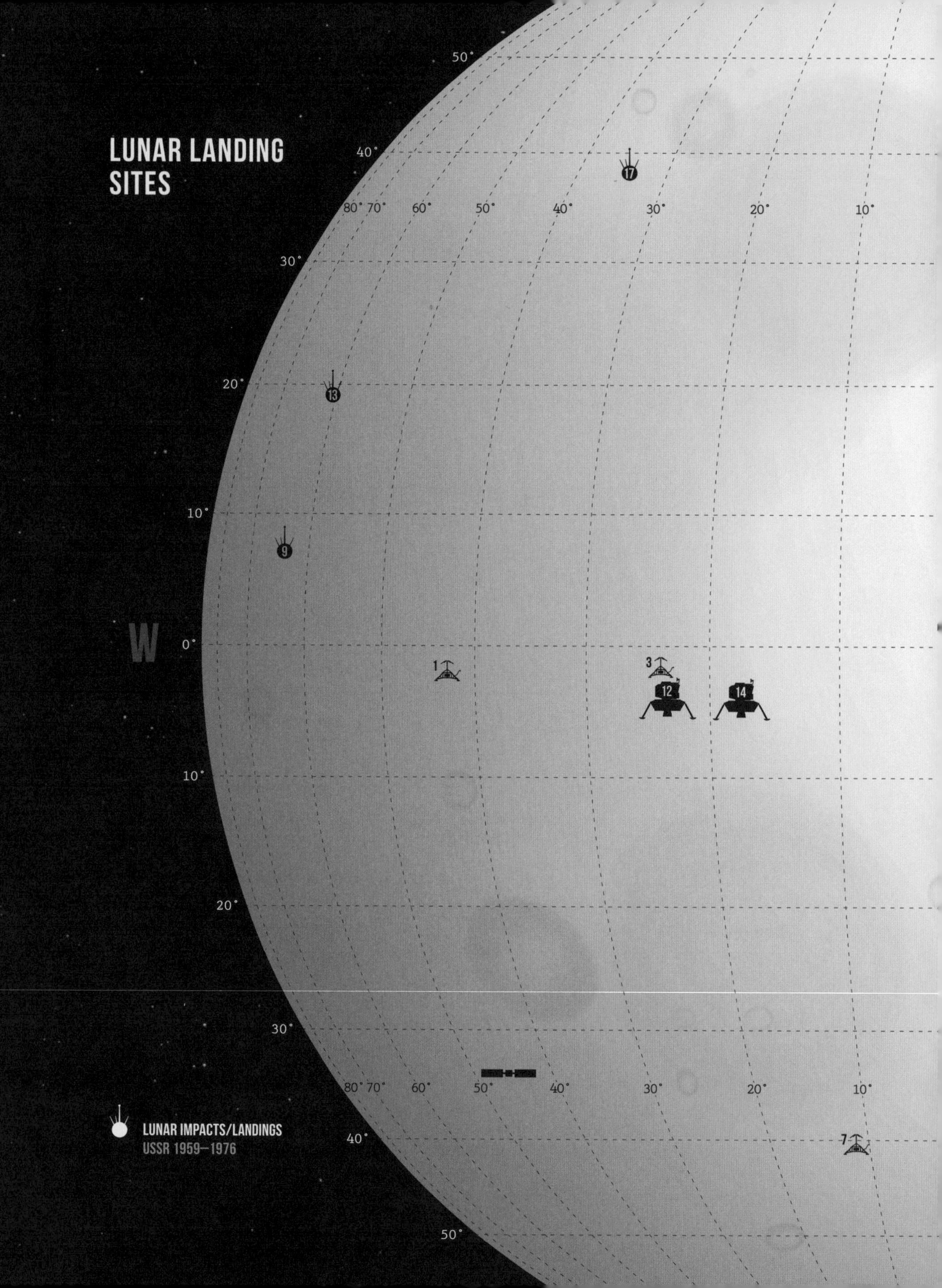
LUNAR LANDING SITES
W
50°
40°
30°
20°
10°
0°
10°
20°
30°
40°
50°
80° 70° 60° 50° 40° 30° 20° 10°
80° 70° 60° 50° 40° 30° 20° 10°
17
13
9
1
3
12
14
7
LUNAR IMPACTS/LANDINGS
USSR 1959–1976

APOLLO MANNED MISSIONS
USA 1969 – 1972
SURVEYOR LANDINGS
USA 1966 – 1968
HITEN IMPACT
JAPAN 1993
SMART - 1
EUROPEAN SPACE AGENCY
2006
E

MISSION INSIGNIA AND SPACECRAFT CALL SIGNS

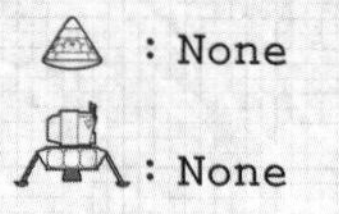

: None

: None

: Apollo 7

: None

: Apollo 8

: None

: Yankee Clipper

: Intrepid

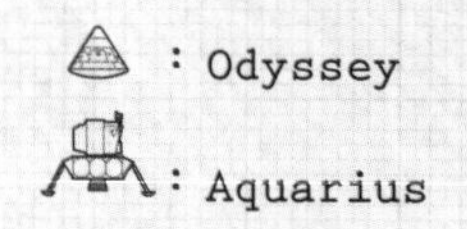

: Odyssey

: Aquarius

: Kitty Hawk

: Antares

For each manned mission, an embroidered patch was created that was worn by the astronauts and personnel associated with the mission.

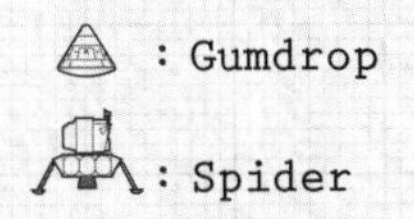
: Gumdrop

: Spider

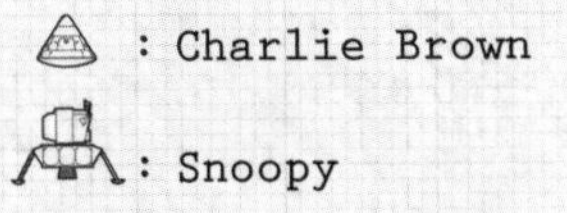
: Charlie Brown

: Snoopy

: Columbia

: Eagle

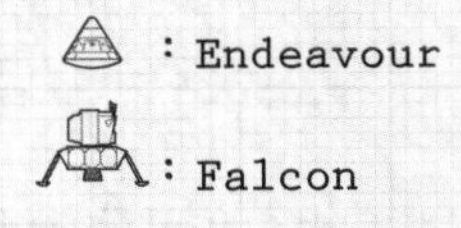
: Endeavour

: Falcon

: Casper

: Orion

: America

: Challenger

On each mission the Command Module and the Lunar Module were given call signs that were used to identify the craft during radio communications.

ASTRONAUT WEIGHTS

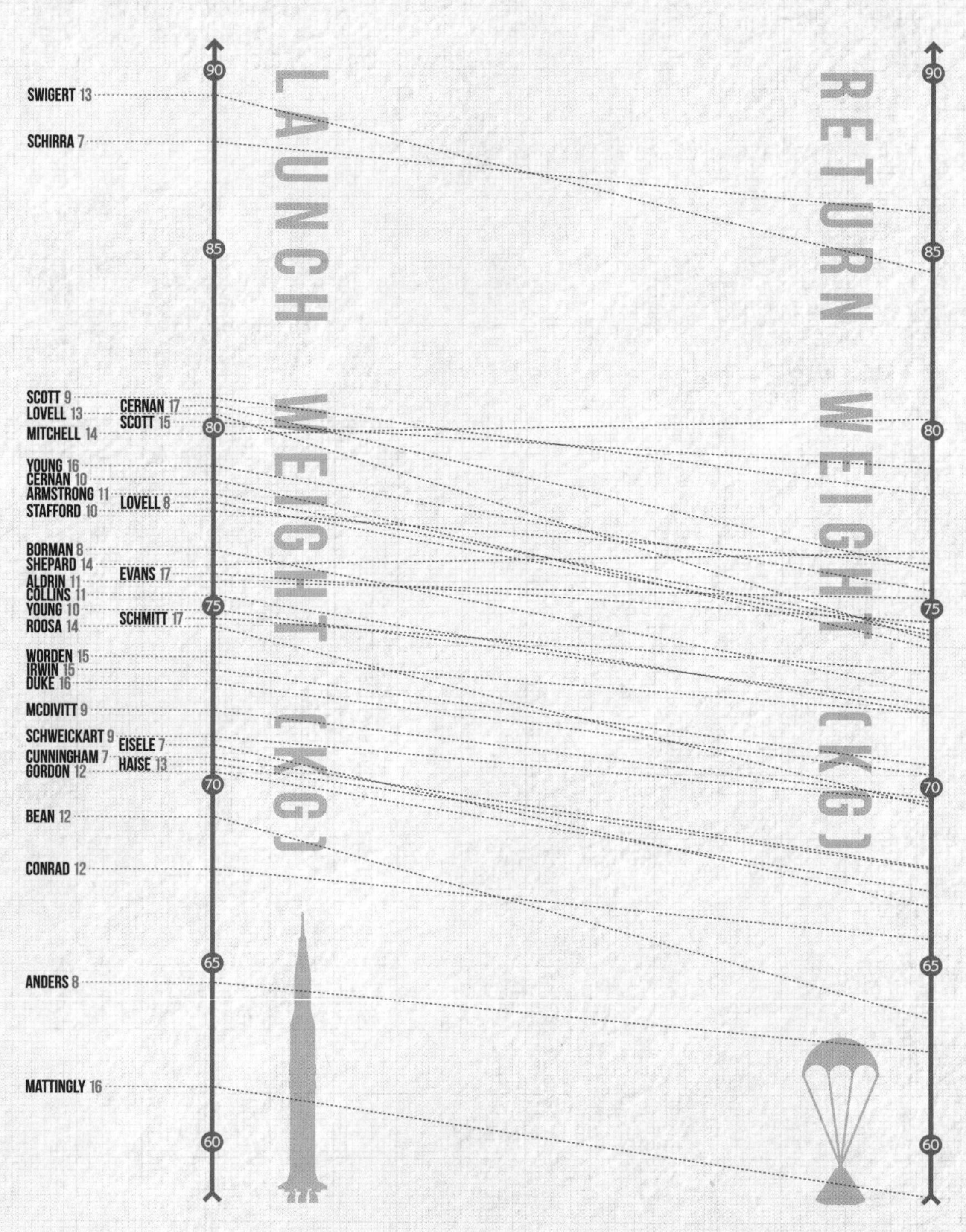

LAUNCH SPECTATORS

For safety reasons, spectators watched lift-off from a distance of at least 5 km.

5KM

LARGEST MOONS IN THE SOLAR SYSTEM

Earth's Moon is the fifth largest in the solar system. Its diameter is more than a quarter of Earth's, making it the largest moon compared to its host planet.

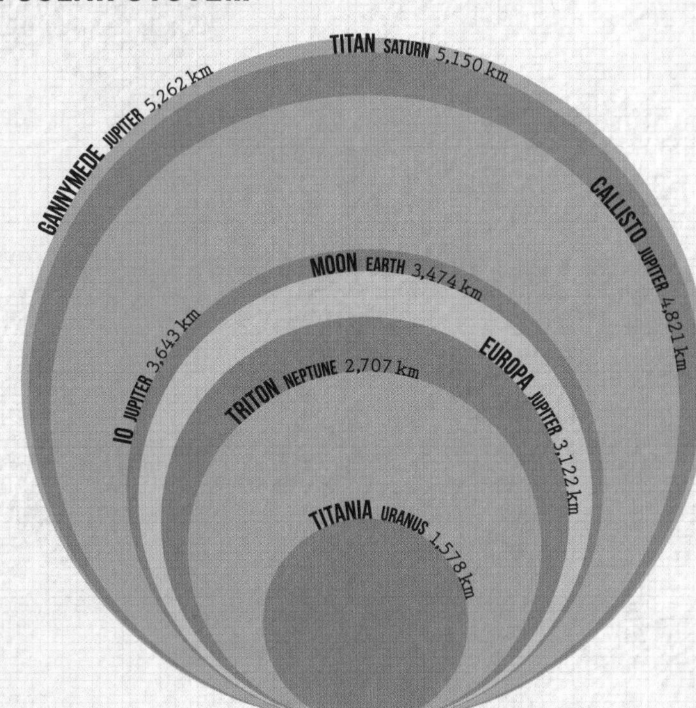

LUNAR MODULE LANDING ACCURACY

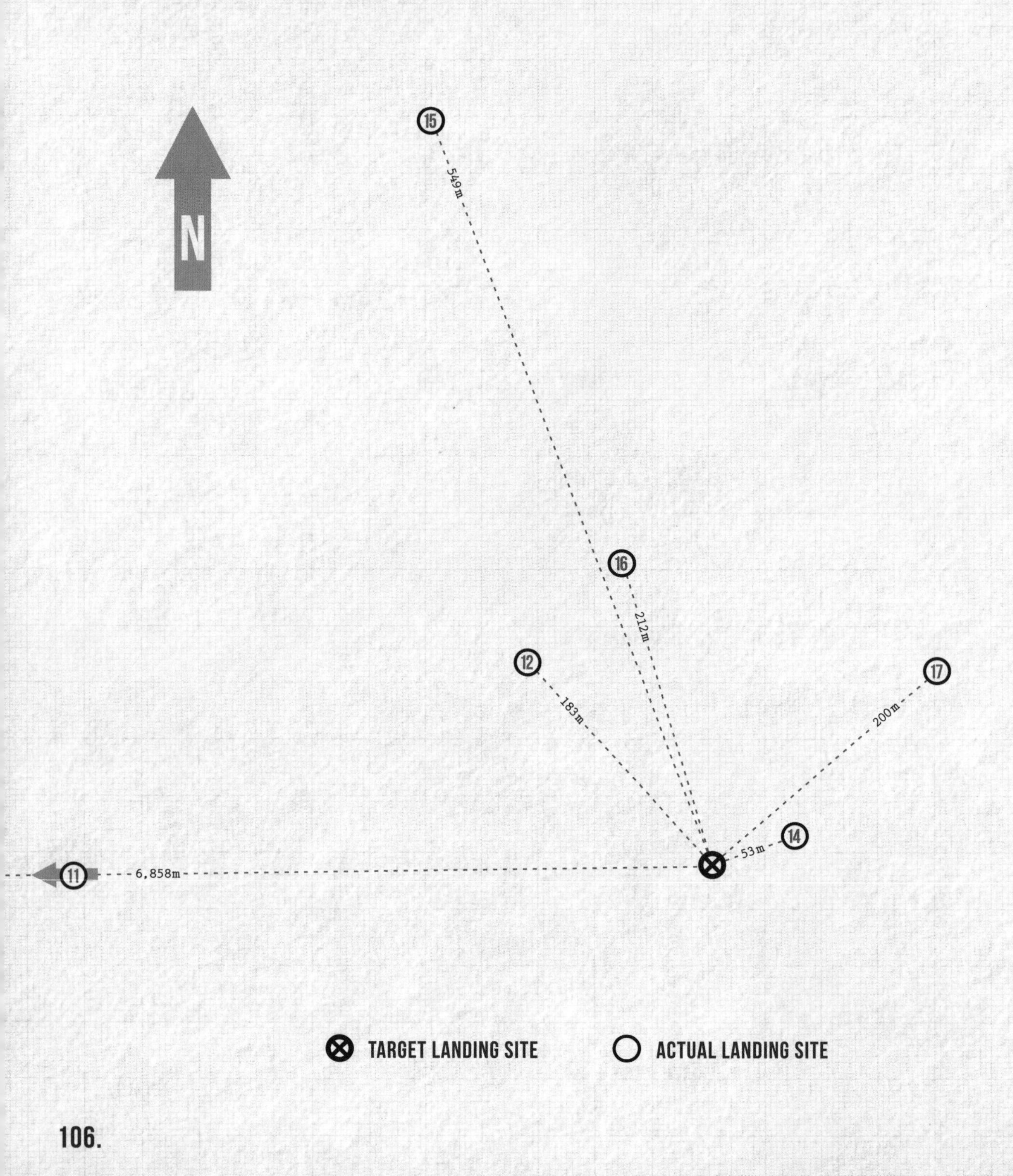

MOON COMPOSITION

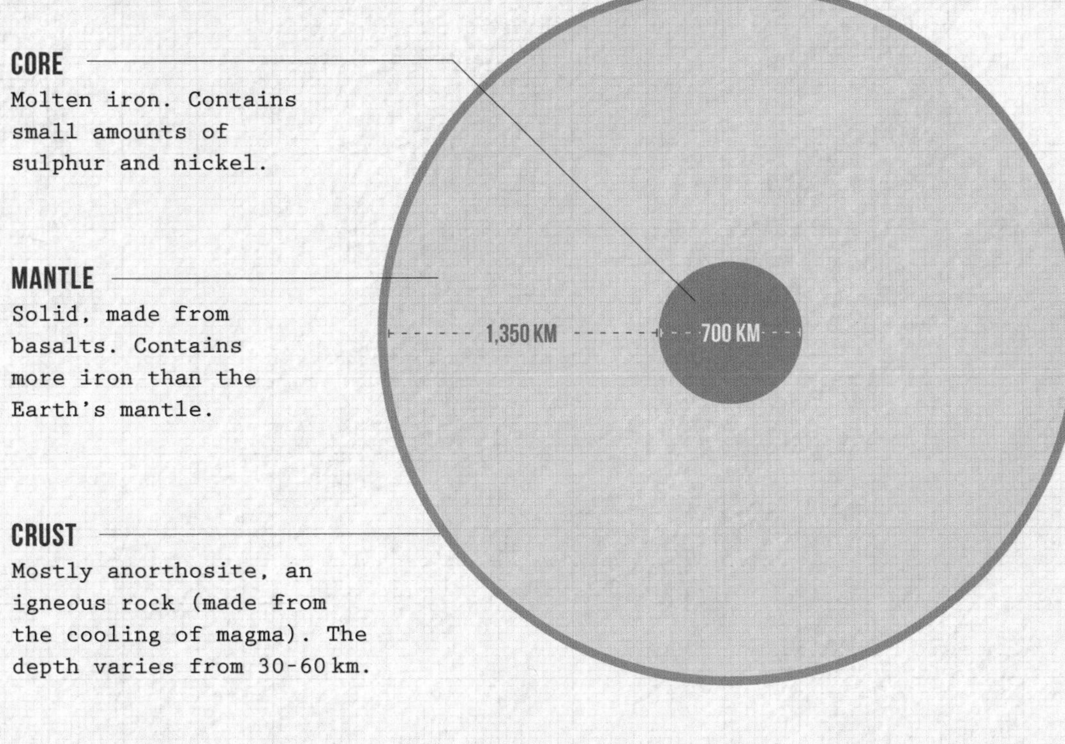

ELEMENTS IN THE LUNAR CRUST

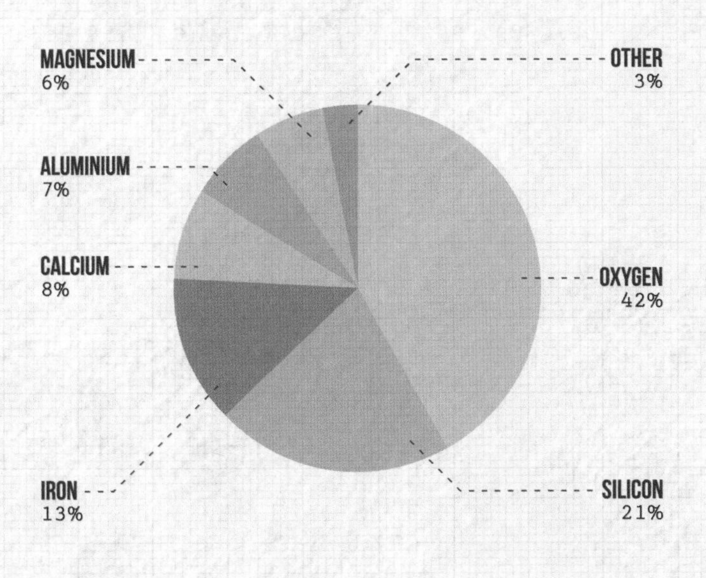

Because of the samples brought back by the Apollo program, scientists have gathered more information about the crust than any other part of the Moon.

The chart shows the average concentration of elements present in the lunar rock and soil.

DISTANCES TRAVELLED BY COMMAND MODULES

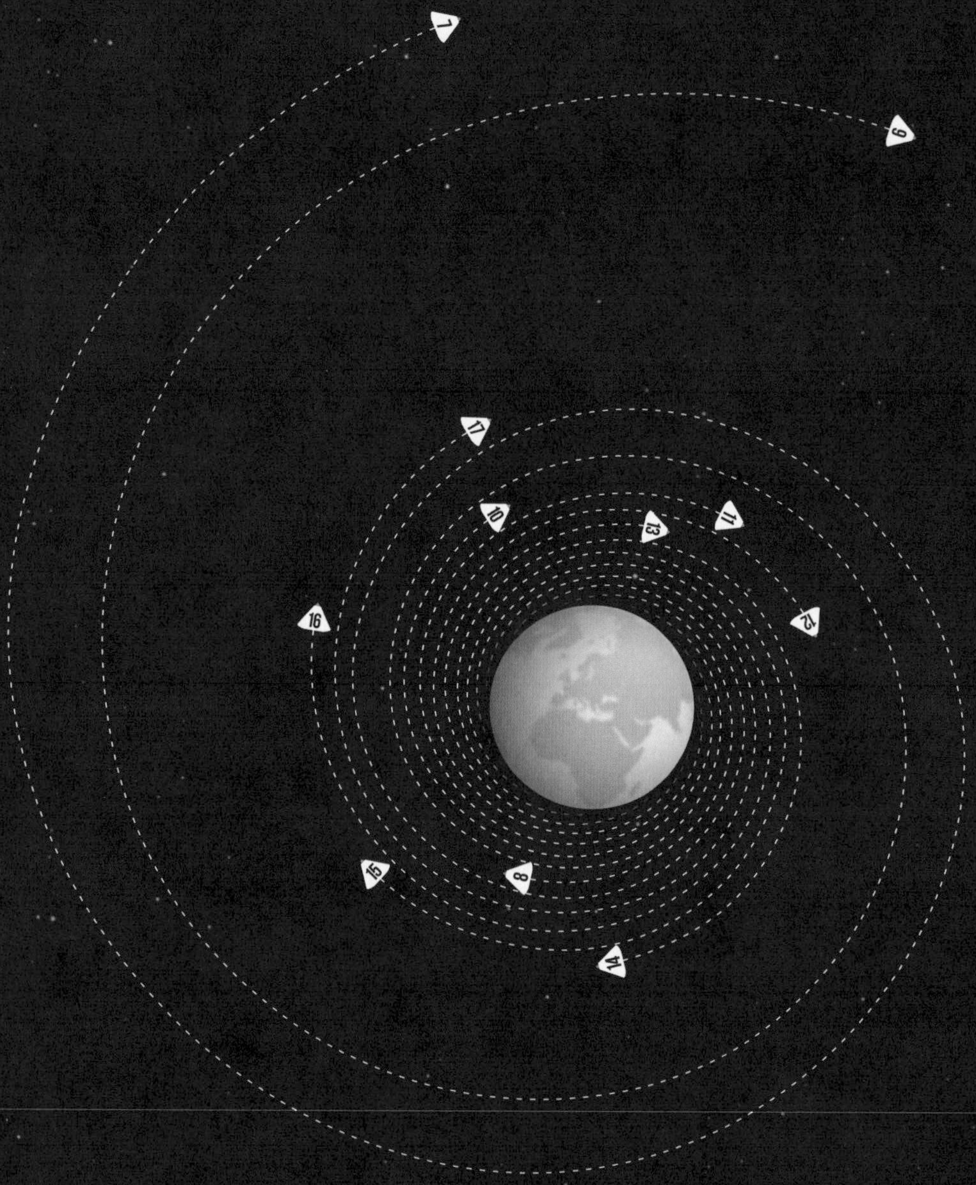

7
9
17
10
13
11
16
12
15
8
14

APOLLO 7:	322,515 km	**APOLLO 11:**	153,4832 km	**APOLLO 15:**	2,051,914 km
APOLLO 8:	933,419 km	**APOLLO 12:**	1533,704 km	**APOLLO 16:**	2,238,598 km
APOLLO 9:	6,787,247 km	**APOLLO 13:**	1,002,123 km	**APOLLO 17:**	2,391,485 km
APOLLO 10:	1,335,755 km	**APOLLO 14:**	1,852,516 km		

SPLASHDOWN – DISTANCE TO TARGET

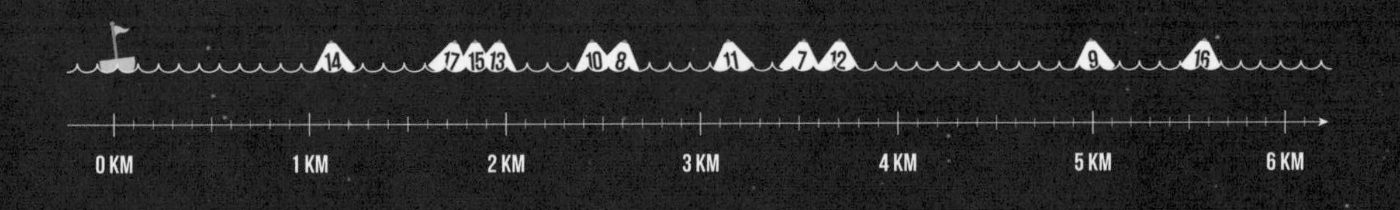

ASTRONAUT AGE

50
40
30
20
10

AGE:
WALTER SCHIRRA
DONN EISELE
WALTER CUNNINGHAM
FRANK BORMAN
JIM LOVELL
WILLIAM ANDERS
JAMES MCDIVITT
DAVID SCOTT
RUSTY SCHWEICKART
THOMAS STAFFORD
JOHN YOUNG
EUGENE CERNAN
NEIL ARMSTRONG
MICHAEL COLLINS
BUZZ ALDRIN
PETE CONRAD
RICHARD GORDON
ALAN BEAN
JACK SWIGERT
FRED HAISE
ALAN SHEPARD
STUART ROOSA
EDGAR MITCHELL
ALFRED WORDEN
JAMES IRWIN
KEN MATTINGLY
CHARLES DUKE
RONALD EVANS
HARRISON SCHMITT

Shown above are the astronauts' ages on their first Apollo launch.

ACCUMULATED TIME IN SPACE

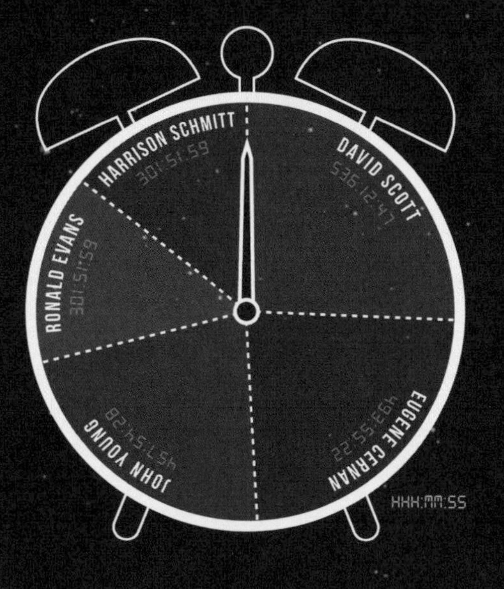

Total time spent in space for the five most travelled Apollo astronauts.

TOP SPEEDS OF LUNAR ROVERS

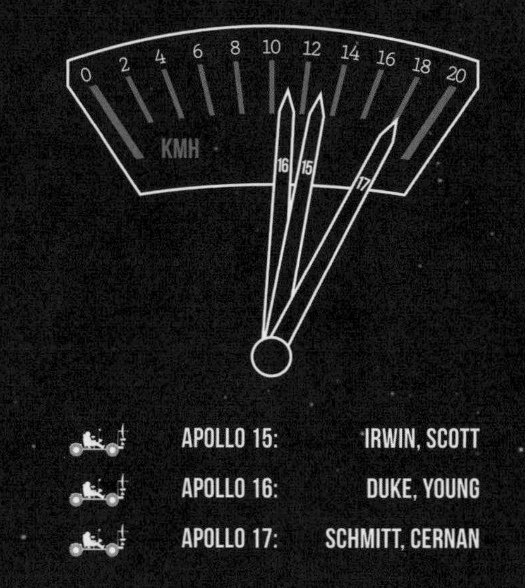

The top speeds achieved by the three LRVs on Apollo missions 15, 16 & 17.

DISTANCE TO THE MOON

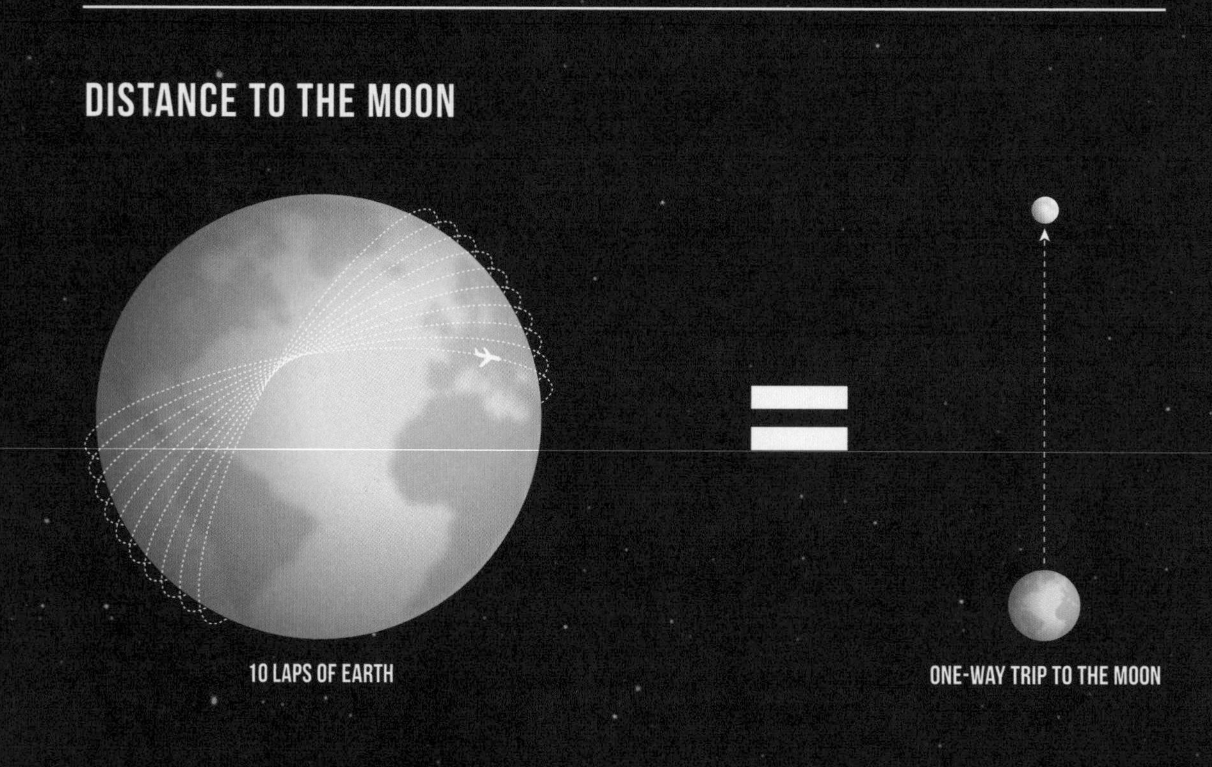

EARTH ORBIT ALTITUDE RANGES

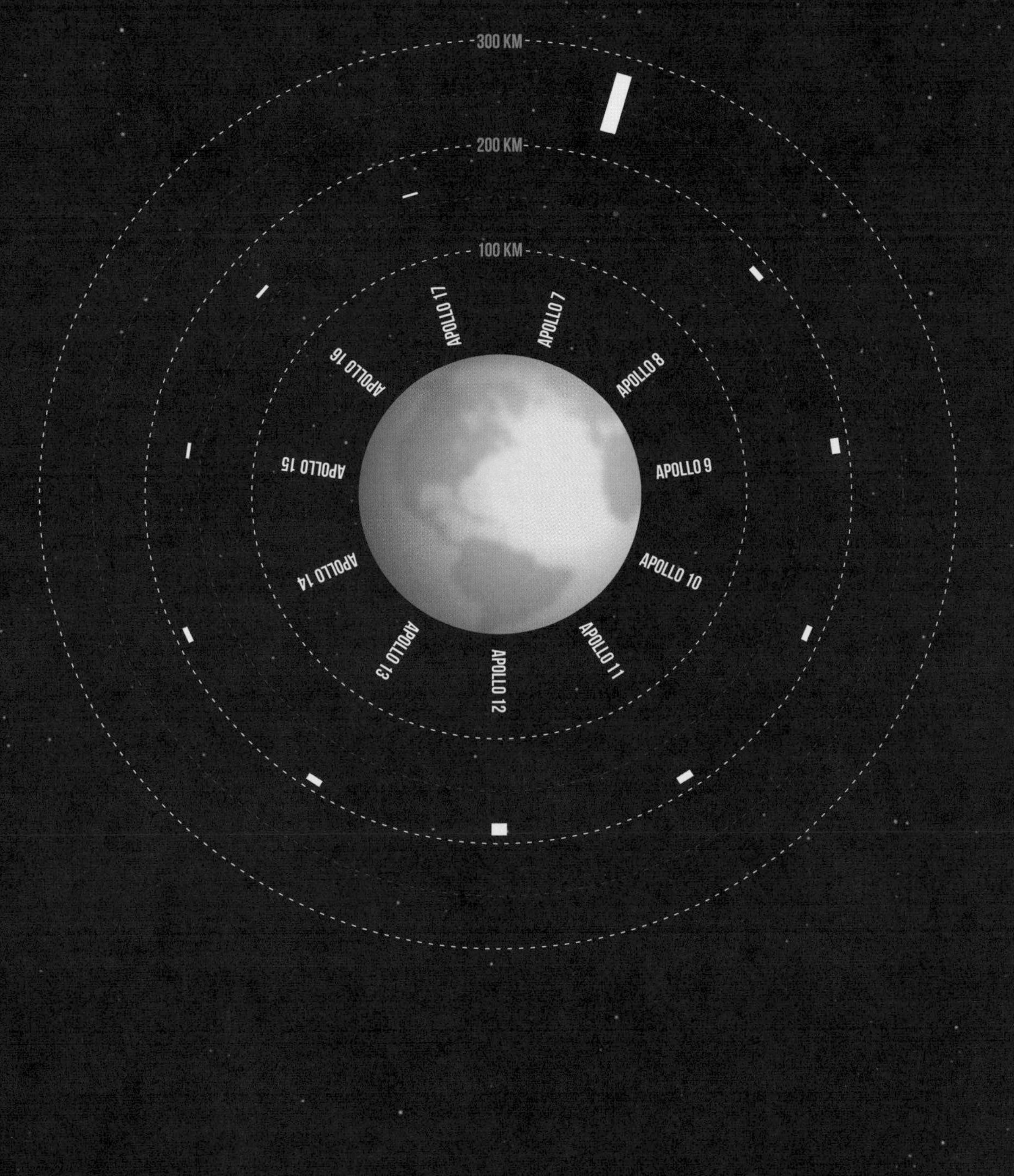

DISTANCES TRAVELLED BY LUNAR ROVER

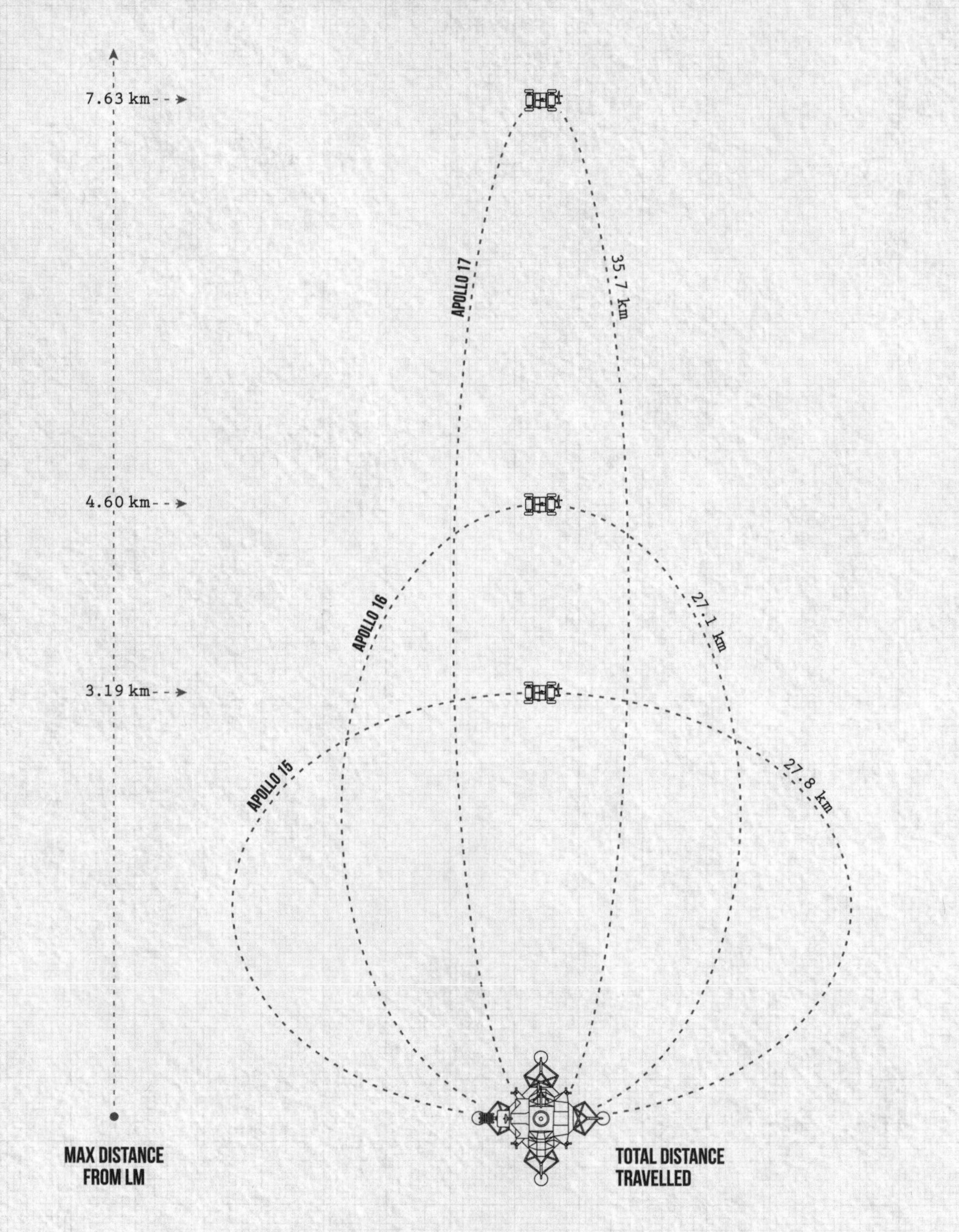

APOLLO FOOD RATIONS

BREAKFAST ITEMS

APRICOT
BACON SQUARES
CANADIAN BACON & APPLESAUCE
CINNAMON TOASTED BREAD CUBES
CORNFLAKES
FRUIT COCKTAIL
PEACHES
SAUSAGE PATTIES
SCRAMBLED EGGS
SPICED FRUIT CEREAL

MEATS

BEEF & GRAVY
BEEF & VEGETABLES
BEEF POT ROAST
BEEF STEW
CHICKEN & RICE
CHICKEN & VEGETABLES
CHICKEN STEW
FRANKFURTERS
MEATBALLS & SAUCE
PORK & SCALLOPED POTATOES
SPAGHETTI & MEAT SAUCE
TURKEY & GRAVY

CUBES & CANDY

BEEF JERKY
BEEF SANDWICHES
BROWNIES
CARAMEL CANDY
CHEESE CRACKERS
CHEESE SANDWICHES
CHOCOLATE BARS
CREAMED CHICKEN BITES
JELLIED FRUIT CANDY
PEANUT CUBES
PECANS
PINEAPPLE FRUITCAKE
SUGAR COOKIES
TURKEY BITES

SALADS AND SOUPS

CHICKEN & RICE SOUP
LOBSTER BISQUE
PEA SOUP
POTATO SOUP
SHRIMP COCKTAIL
TOMATO SOUP
TUNA SALAD

SANDWICH SPREADS & BREAD

BREAD
CHEDDAR CHEESE
CHICKEN SALAD
HAM SALAD
JELLY
KETCHUP
MUSTARD
PEANUT BUTTER

DESSERTS

APPLESAUCE
BANANA PUDDING
BUTTERSCOTCH PUDDING
CHOCOLATE PUDDING
CRANBERRY-ORANGE SAUCE
PEACH AMBROSIA

BEVERAGES

COCOA
COFFEE
GRAPE DRINK
GRAPE PUNCH
GRAPEFRUIT DRINK
ORANGE-GRAPEFRUIT DRINK
ORANGE JUICE
PINEAPPLE-GRAPEFRUIT DRINK
PINEAPPLE-ORANGE DRINK

RS – REHYDRATABLE SPOON BOWL

A plastic container whose contents were eaten with a spoon.

RD – REHYDRATABLE DRINK

Dry ingredients that were mixed with water inside a ziplock bag.

IM – INTERMEDIATE MOISTURE

Low-moisture foods with additives to prevent the growth of molds and microorganisms.

D – DEHYDRATED

Dry ingredients that were mixed with water inside ziplock bags.

T – THERMOSTABILIZED

Foods that were heat processed to destroy harmful microorganisms.

NS – NATURAL STATE

Foods like nuts and cereal bars that were sealed in clear pouches.

SATURN V HEIGHT

SATURN V 111 m

BIG BEN 96 m

STATUE OF LIBERTY 93 m

BOEING 747 71 m

SPACE SHUTTLE 56 m

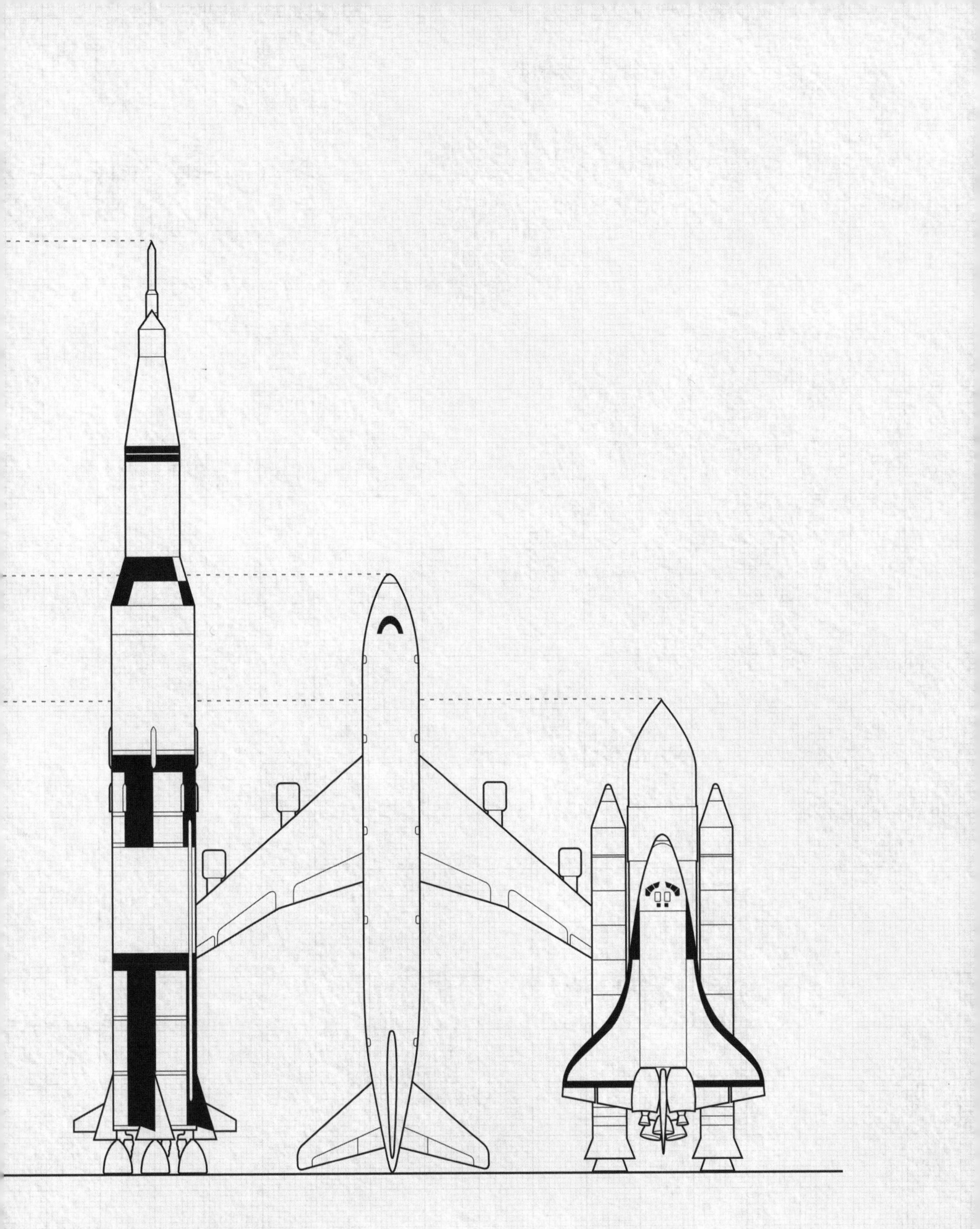

COMMAND MODULE CONTROL PANEL

The main display was more than 2m wide and 91cm tall. It was formed of three panels that displayed information and controls specific to each crew member.

71 LIGHTS

40 MECHANICAL INDICATORS

24 INSTRUMENTS

566 SWITCHES

COMMAND MODULE WIRING

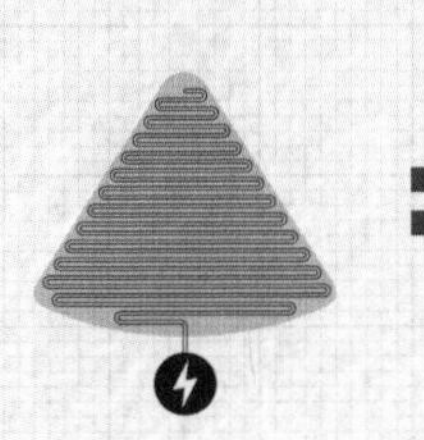

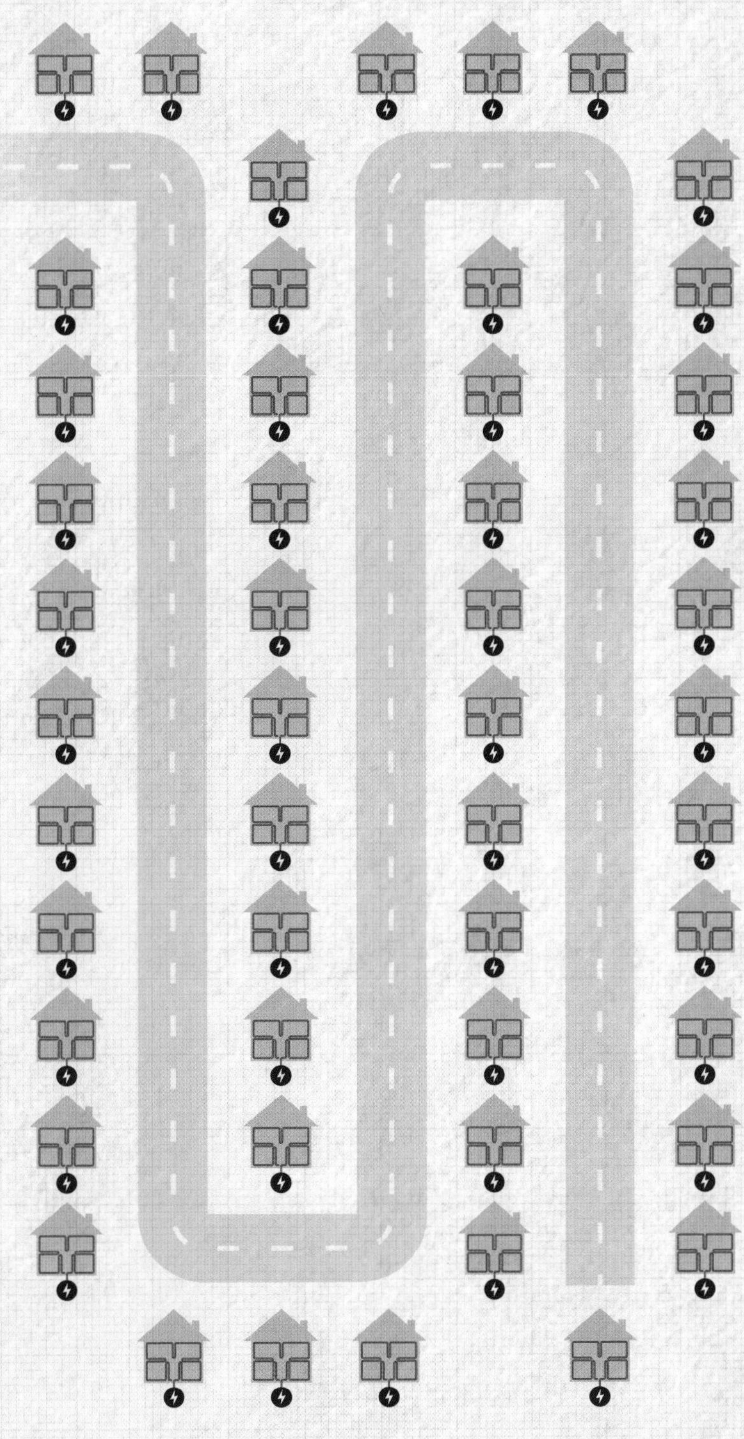

There were 24km of wiring in the Command Module, roughly the amount needed to wire fifty two-bedroom homes.

APOLLO 14 CSM SPEED COMPARISON

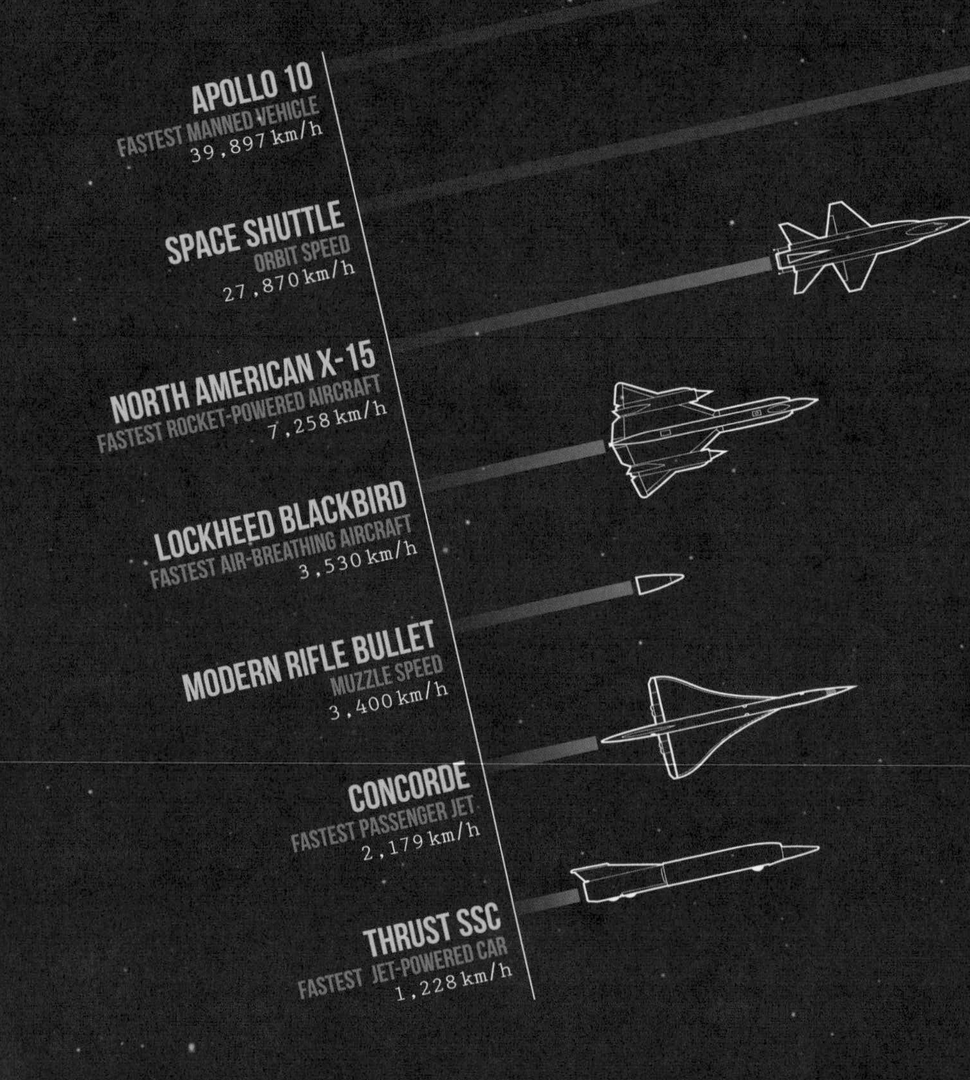

The CSM of Apollo 10 reached the highest speed of the program. The crew still holds the record of being the fastest men in history.

EARTH SPLASHDOWN SITES

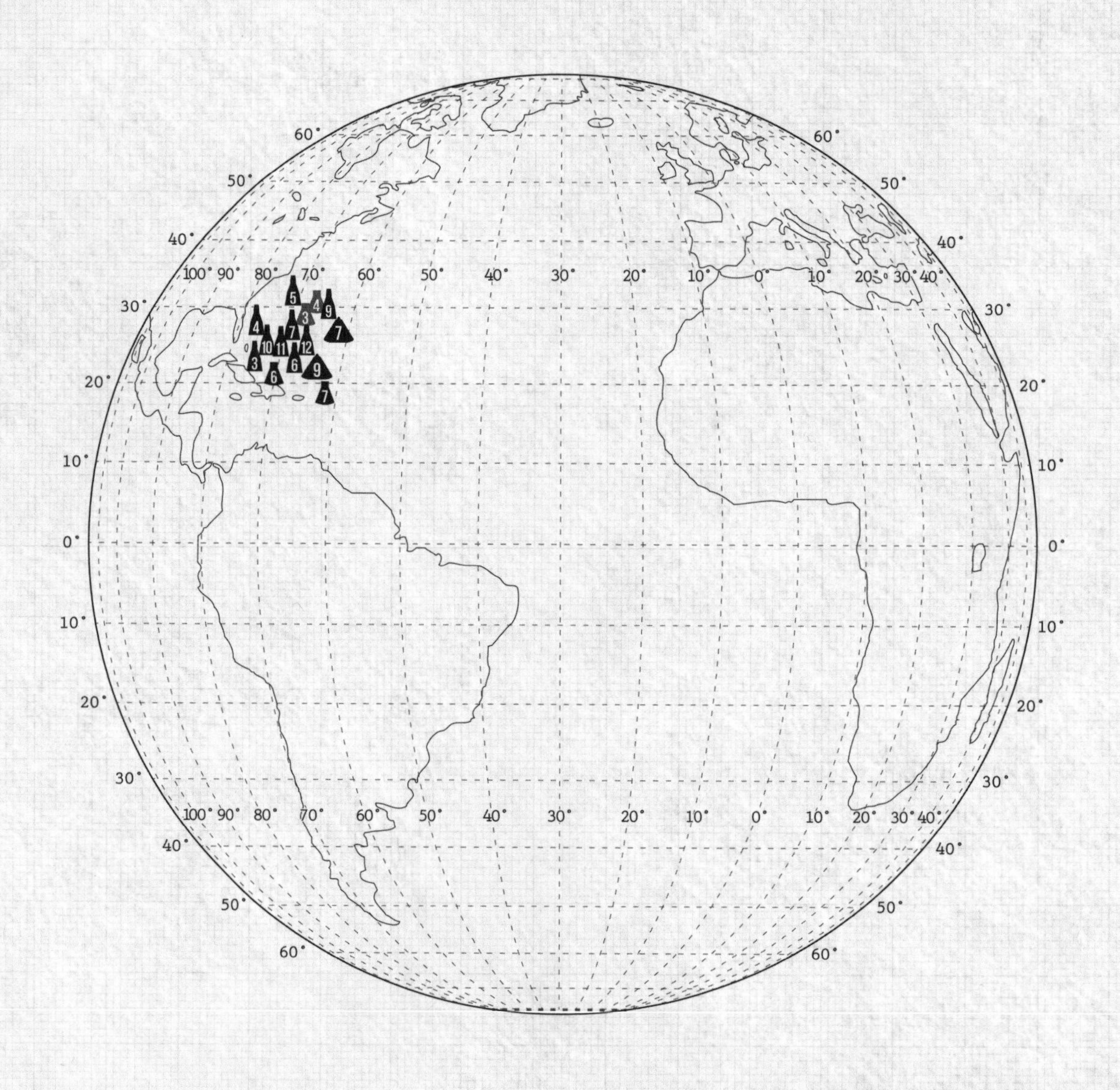

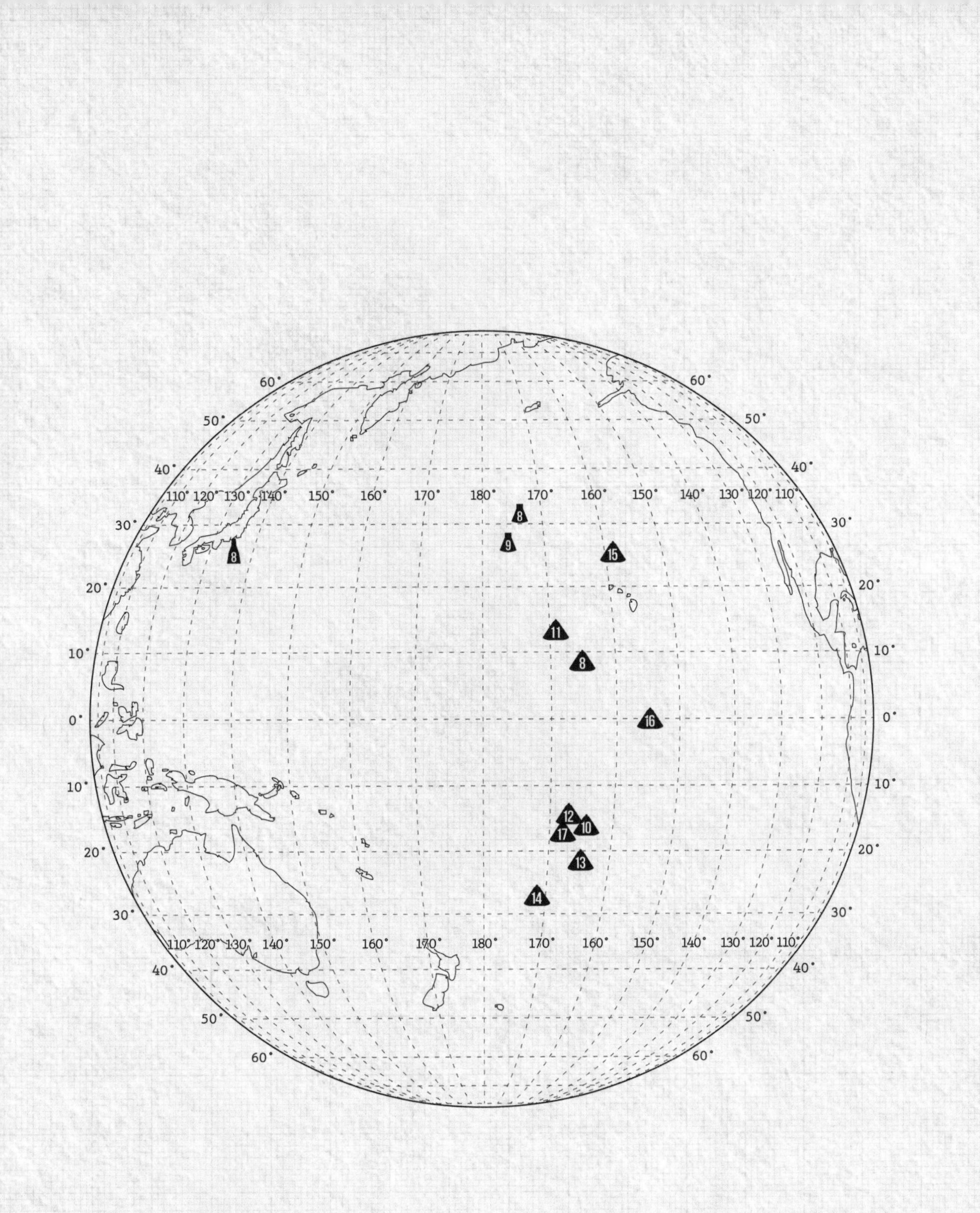

MERCURY-REDSTONE SPLASHDOWN
MERCURY-ATLAS SPLASHDOWN

SATURN V FUEL CONSUMPTION

From lift-off, and for the duration of the Saturn V's first-stage burn, the five F1 engines burned roughly 12,000 kg of propellant every second.

This is the same amount of fuel as 1,500 fighter jets using afterburn. Jets use afterburn to achieve maximum speed, which consumes the most fuel.

APOLLO ASTRONAUTS' BIRTH LOCATIONS

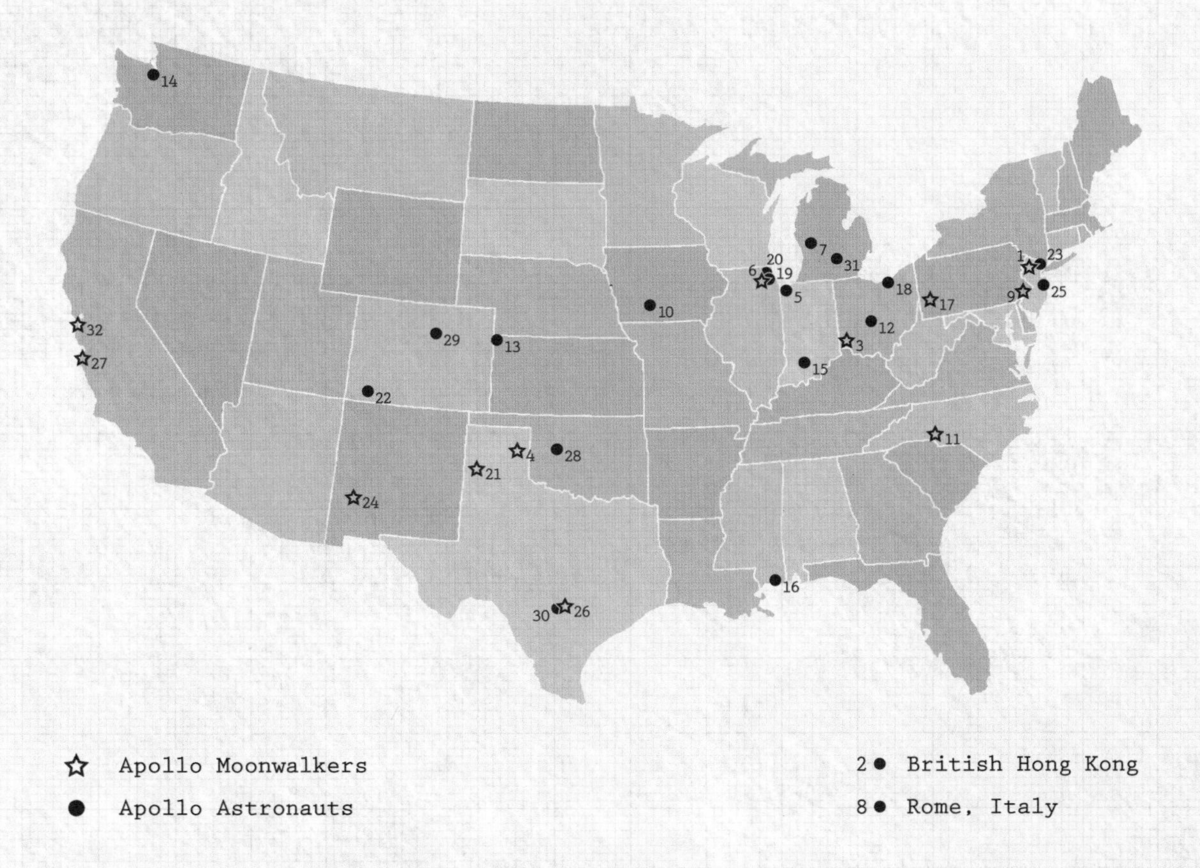

1. Buzz Aldrin
2. William Anders
3. Neil Armstrong
4. Alan Bean
5. Frank Borman
6. Eugene Cernan
7. Roger Chaffee
8. Michael Collins
9. Pete Conrad
10. Walter Cunningham
11. Charles Duke
12. Donn Eisele
13. Ronald Evans
14. Richard Gordon
15. Virgil Grissom
16. Fred Haise
17. James Irwin
18. Jim Lovell
19. Ken Mattingly
20. James McDivitt
21. Edgar Mitchell
22. Stuart Roosa
23. Walter Schirra
24. Harrison Schmitt
25. Rusty Schweickart
26. David Scott
27. Alan Shepard
28. Thomas Stafford
29. Jack Swigert
30. Edward White
31. Alfred Worden
32. John Young

EARTH AND MOON TO SCALE

EARTH DIAMETER
12,742 KM

THE MOON'S FORMATION
20,000 KM - 30,000 KM

It is estimated that the Moon was formed about 4.51 billion years ago, shortly after the creation of our solar system. A planet-sized object crashed into Earth, flinging material into orbit which then coalesced to form the Moon. It has been drifting away from Earth ever since, currently at a rate of about 3.8 cm a year.

GEOSTATIONARY ORBIT
35,786 KM

A satellite orbiting Earth at this distance above the equator, that follows the Earth's rotation, will remain in a fixed position in the sky.

VAN ALLEN RADIATION BELTS

Due to Earth's magnetic field, charged particles are concentrated in these regions. They protect Earth's atmosphere from being destroyed by solar winds and cosmic rays. Their exact size fluctuates depending on the Sun's activity.

INNER BELT
1,000 KM - 6,000 KM

OUTER BELT
13,000 KM - 60,000 KM

MEAN DISTANCE TO MOON
384,400 KM

MOON PERIGEE
363,360 KM

MOON APOGEE
405,500 KM

The exosphere is the boundary between the atmosphere and interplanetary space. Molecules thin out and merge with the vacuum of space.

690 KM

≈

Although considered part of Earth's atmosphere, the air density in the thermosphere is so low that most of it is regarded as outer space. Temperature can fluctuate greatly here, depending on solar activity.

80 KM

Temperature decreases with altitude throughout the mesosphere, with its uppermost region being the coldest part of the atmosphere.

50 KM

The stratosphere holds roughly one-fifth of the Earth's atmosphere. Unlike the troposphere, temperature rises with altitude in the stratosphere. This is due to the ozone layer absorbing the Sun's radiation.

17 KM

The lowest level of the atmosphere, the troposphere, contains more than three-quarters of the mass of Earth's atmosphere. Nearly all weather and cloud formations take place here.

THE SOLAR SYSTEM'S OTHER MOONS

MERCURY: 0

VENUS: 0

EARTH: 1

MARS: 2

JUPITER: 69

SATURN: 62

URANUS: 27

NEPTUNE: 14

PLUTO: 5

ASTRONAUTS ABOARD MANNED MISSIONS

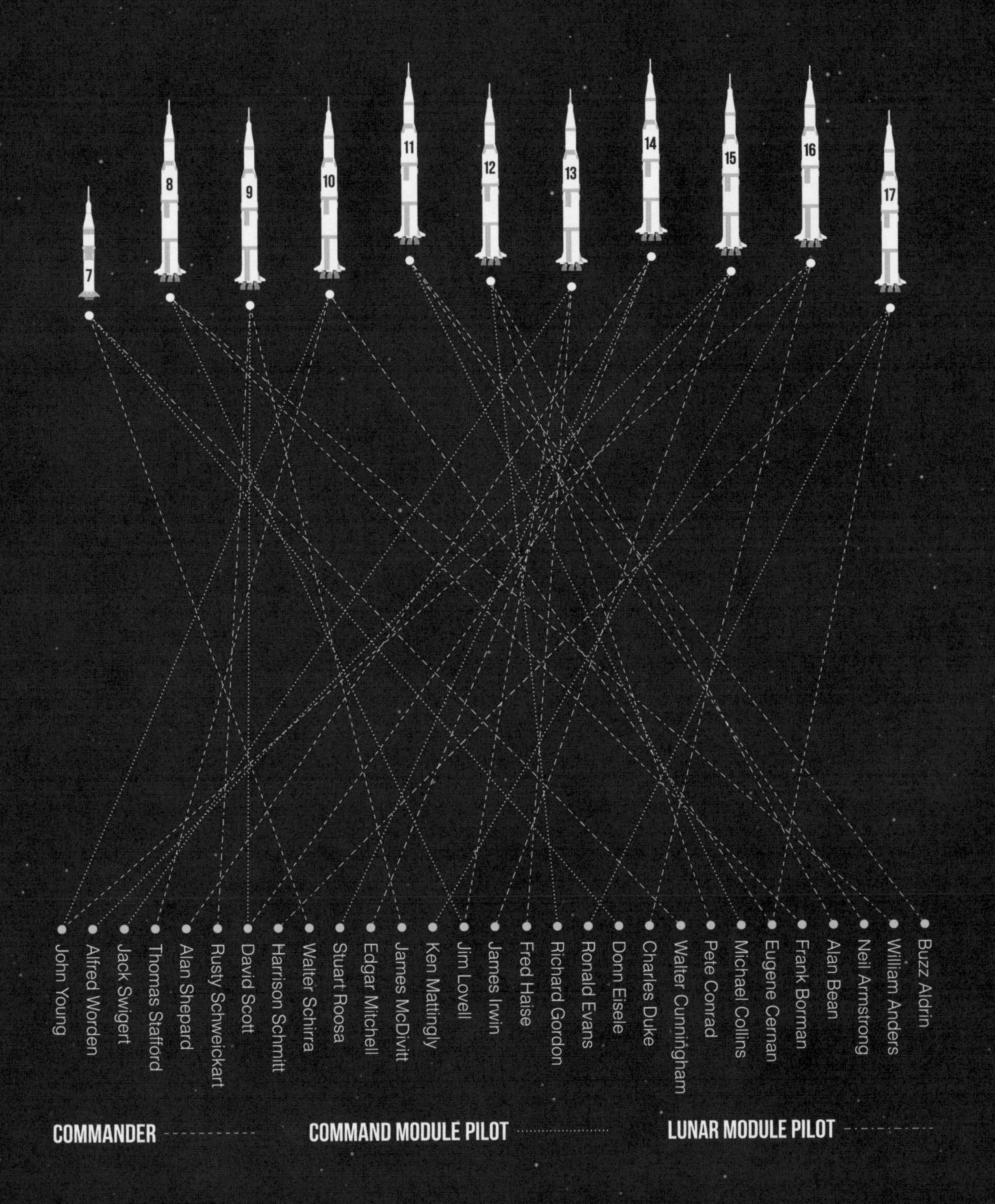

NASA ORGANIZATION CHART

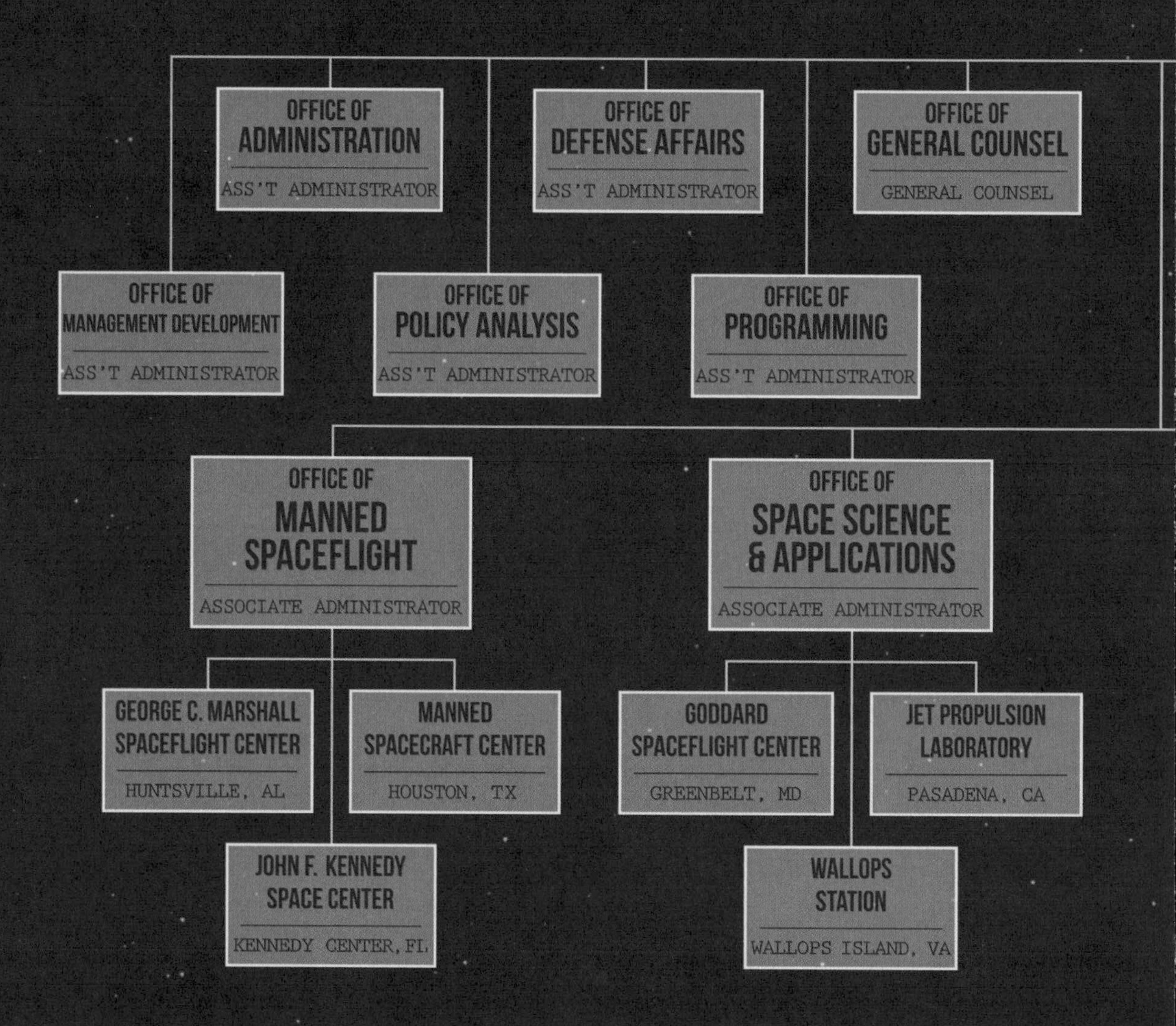

JANUARY 1966

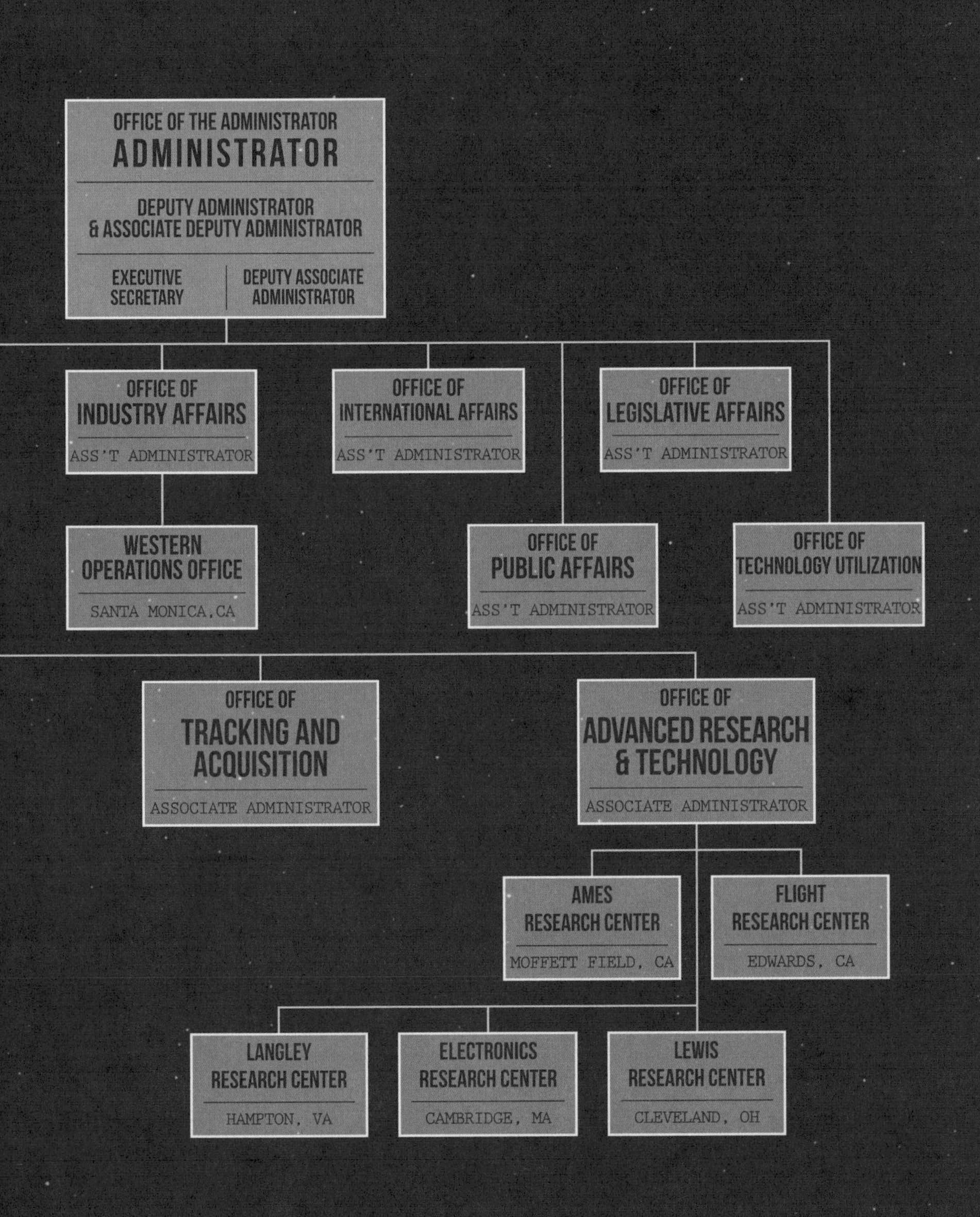
OFFICE OF THE ADMINISTRATOR
ADMINISTRATOR
DEPUTY ADMINISTRATOR
& ASSOCIATE DEPUTY ADMINISTRATOR
EXECUTIVE SECRETARY
DEPUTY ASSOCIATE ADMINISTRATOR
OFFICE OF INDUSTRY AFFAIRS
ASS'T ADMINISTRATOR
OFFICE OF INTERNATIONAL AFFAIRS
ASS'T ADMINISTRATOR
OFFICE OF LEGISLATIVE AFFAIRS
ASS'T ADMINISTRATOR
WESTERN OPERATIONS OFFICE
SANTA MONICA, CA
OFFICE OF PUBLIC AFFAIRS
ASS'T ADMINISTRATOR
OFFICE OF TECHNOLOGY UTILIZATION
ASS'T ADMINISTRATOR
OFFICE OF TRACKING AND ACQUISITION
ASSOCIATE ADMINISTRATOR
OFFICE OF ADVANCED RESEARCH & TECHNOLOGY
ASSOCIATE ADMINISTRATOR
AMES RESEARCH CENTER
MOFFETT FIELD, CA
FLIGHT RESEARCH CENTER
EDWARDS, CA
LANGLEY RESEARCH CENTER
HAMPTON, VA
ELECTRONICS RESEARCH CENTER
CAMBRIDGE, MA
LEWIS RESEARCH CENTER
CLEVELAND, OH

NASA BUDGET

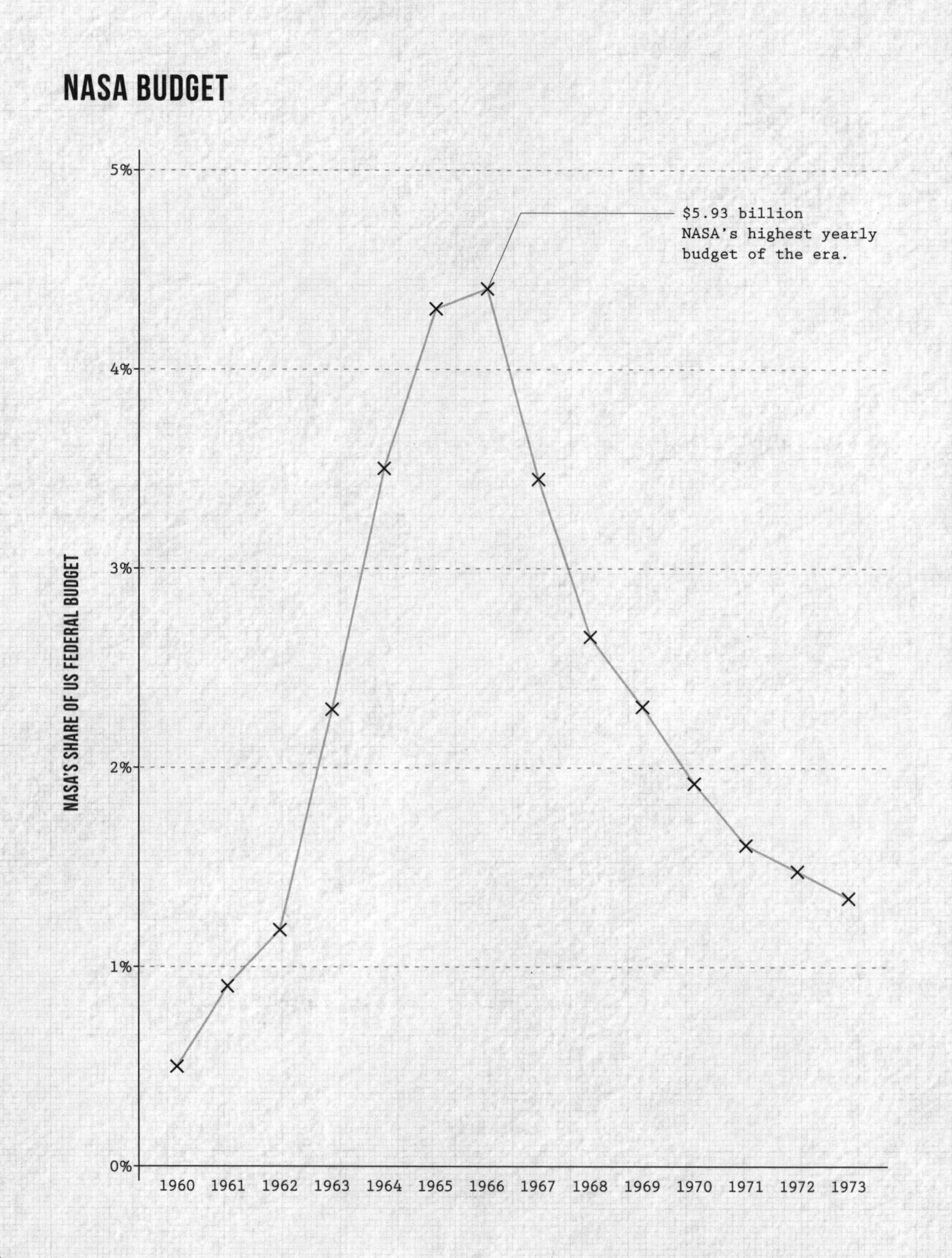

5%
4%
3%
2%
1%
0%
NASA'S SHARE OF US FEDERAL BUDGET
$5.93 billion
NASA's highest yearly
budget of the era.
1960
1961
1962
1963
1964
1965
1966
1967
1968
1969
1970
1971
1972
1973

APOLLO BUDGET

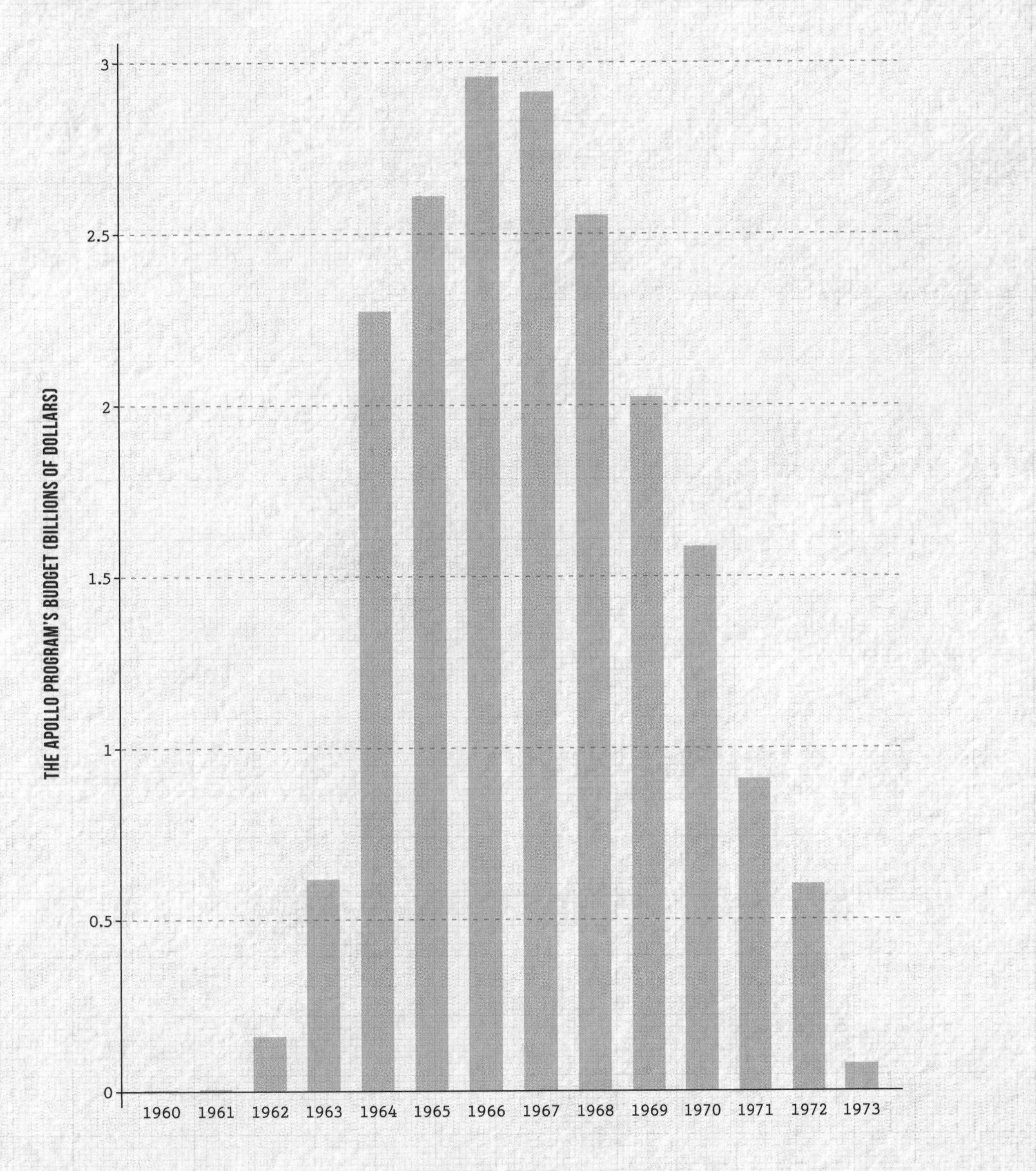

NASA EMPLOYEES

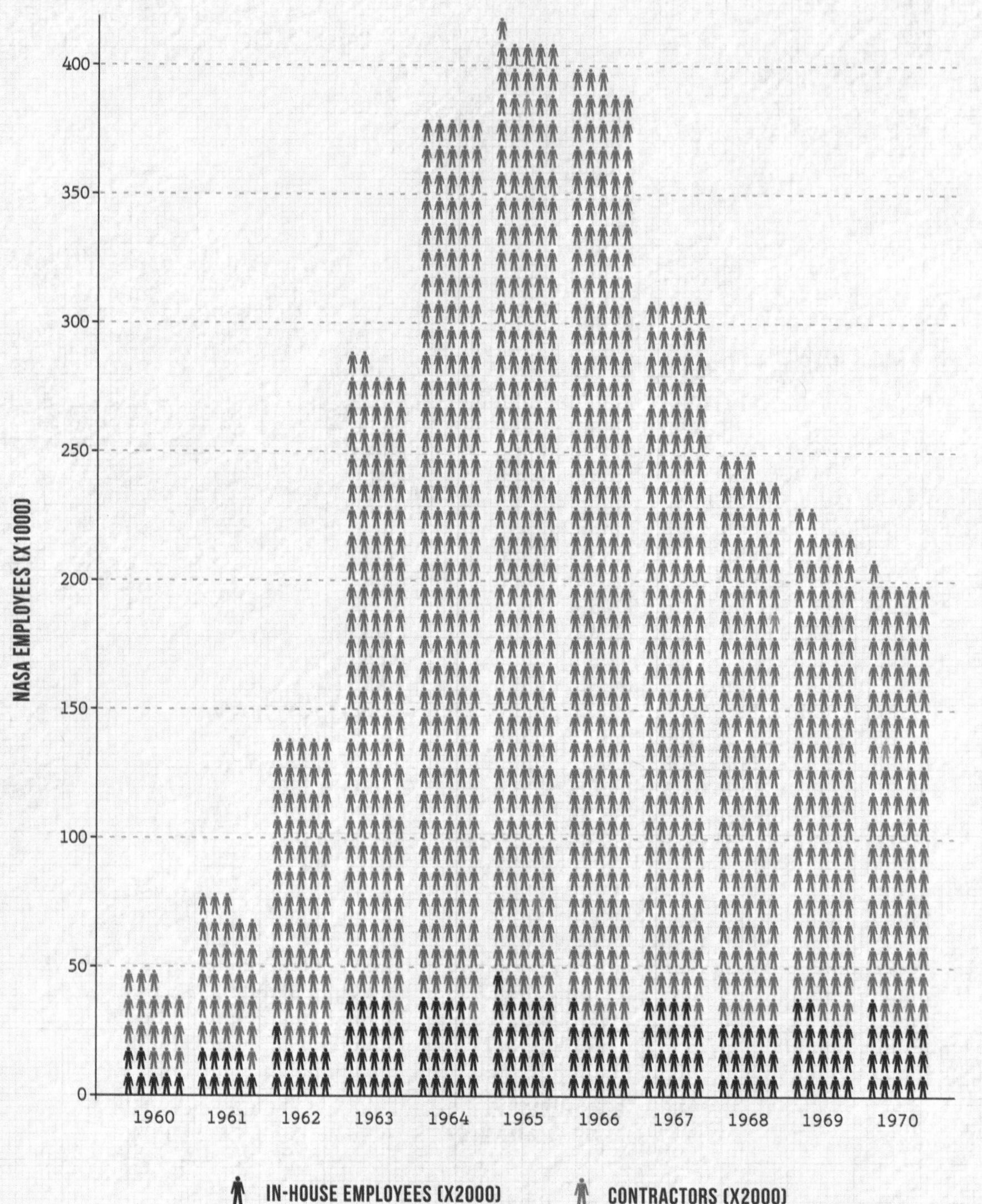

ENGINE POWER

RELATIVE MASS OF APOLLO/SATURN V COMPONENTS

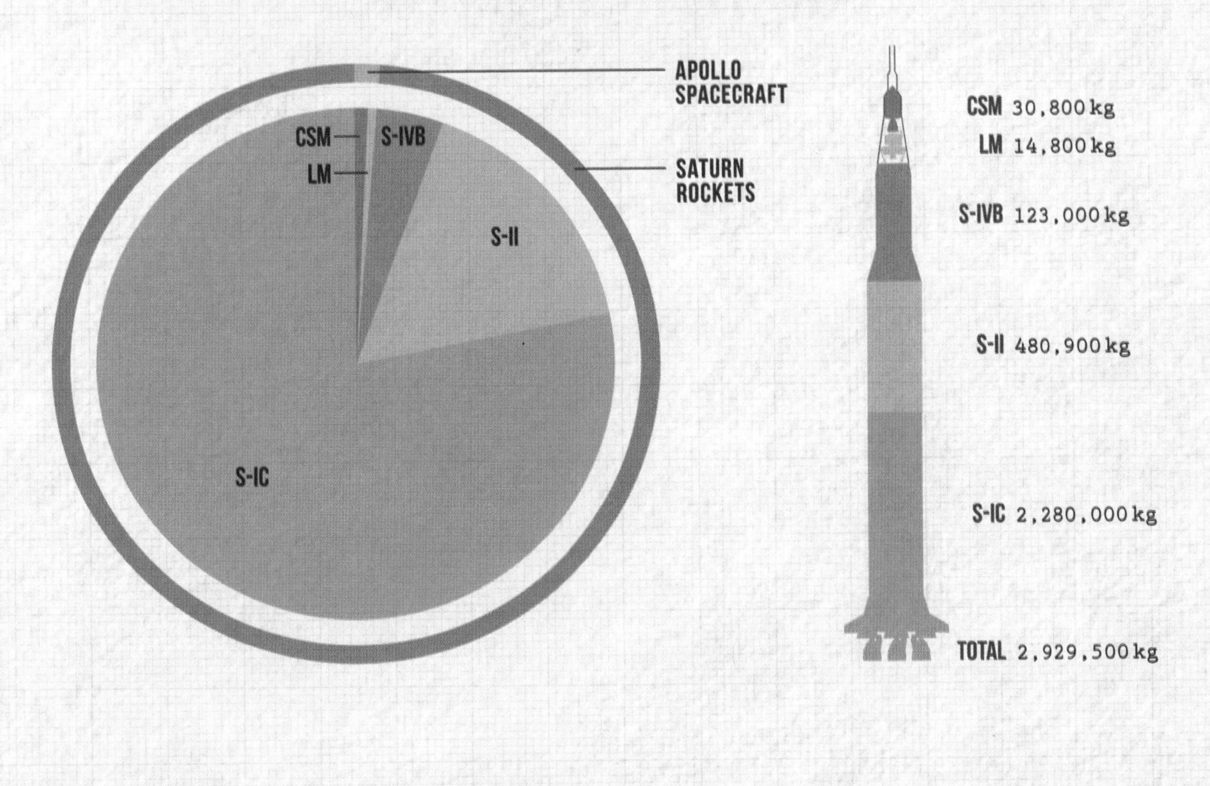

CRAWLER-TRANSPORTER SPEED COMPARISON

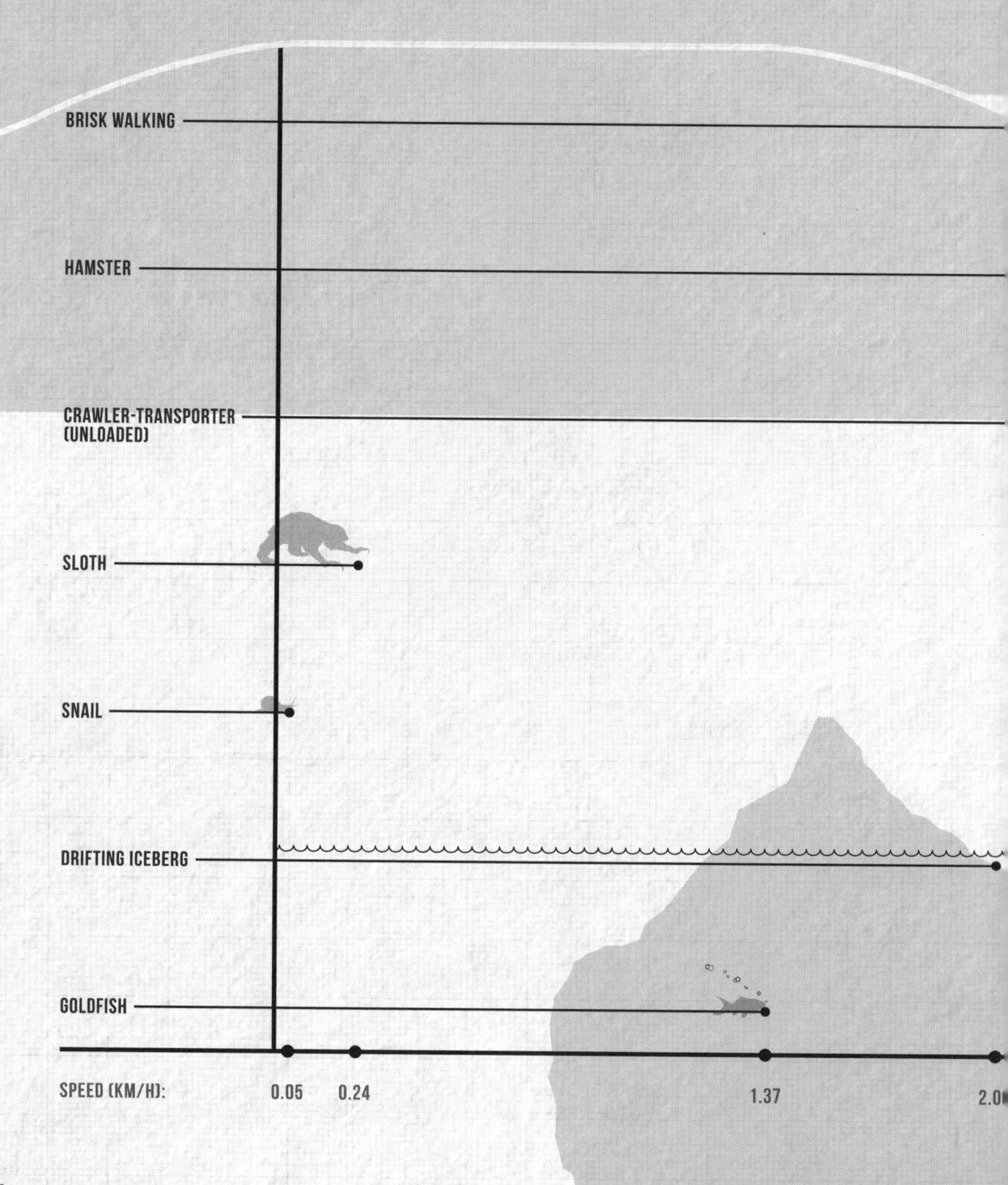

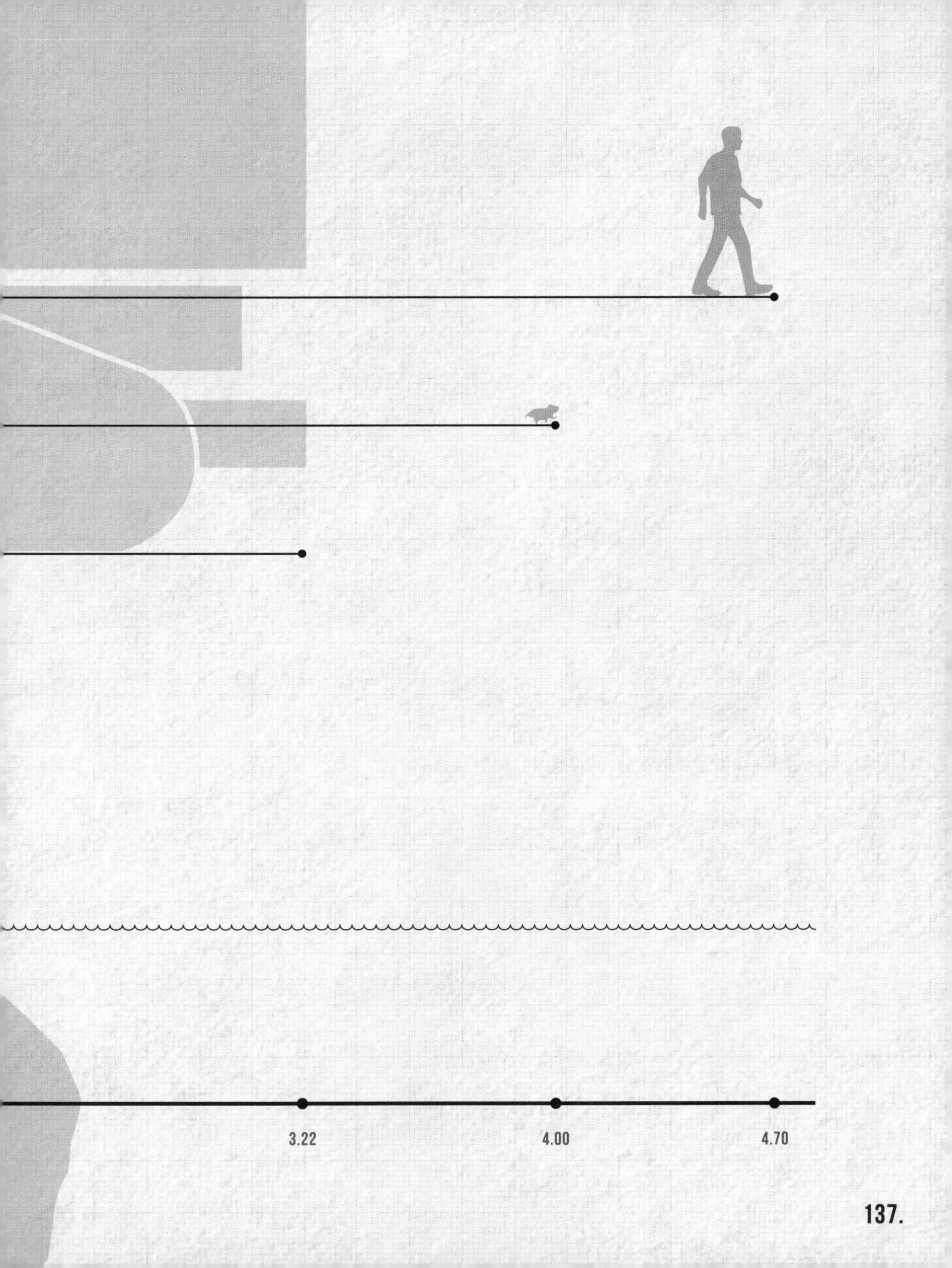
3.22
4.00
4.70

CRAWLER-TRANSPORTER TRACK SIZE

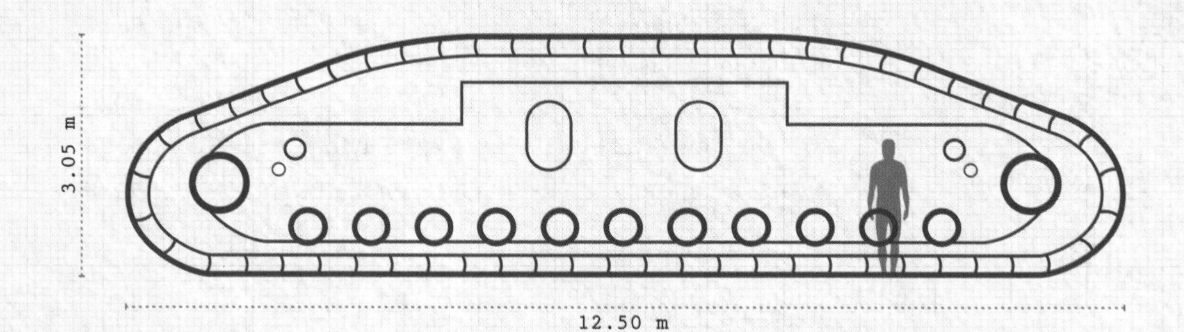

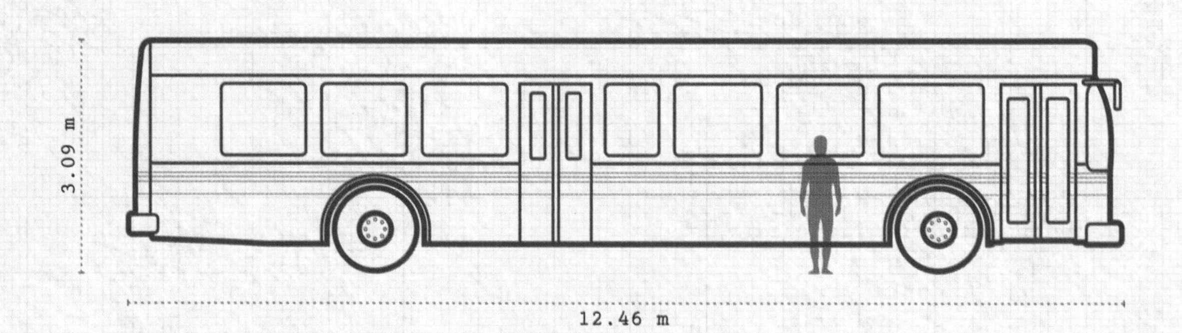

CRAWLER-TRANSPORTER TRACK SHOE MASS

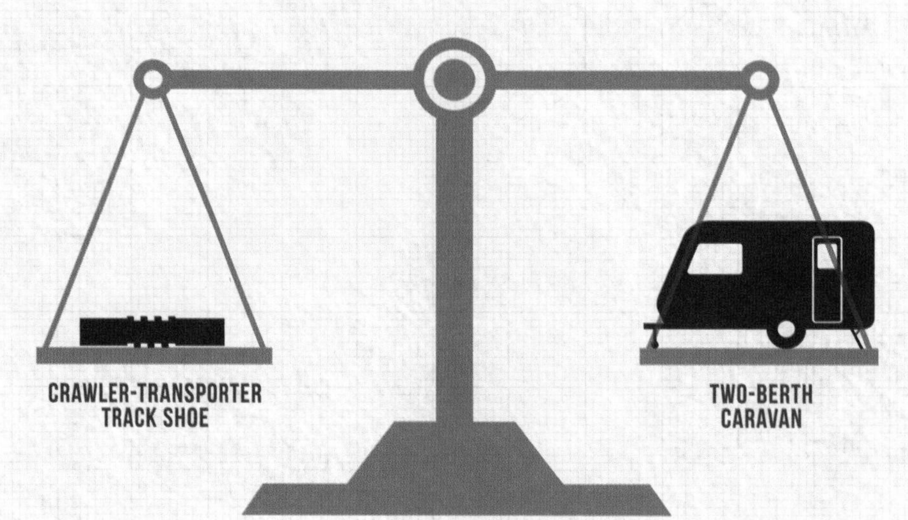

CRAWLER-TRANSPORTER FUEL EFFICIENCY

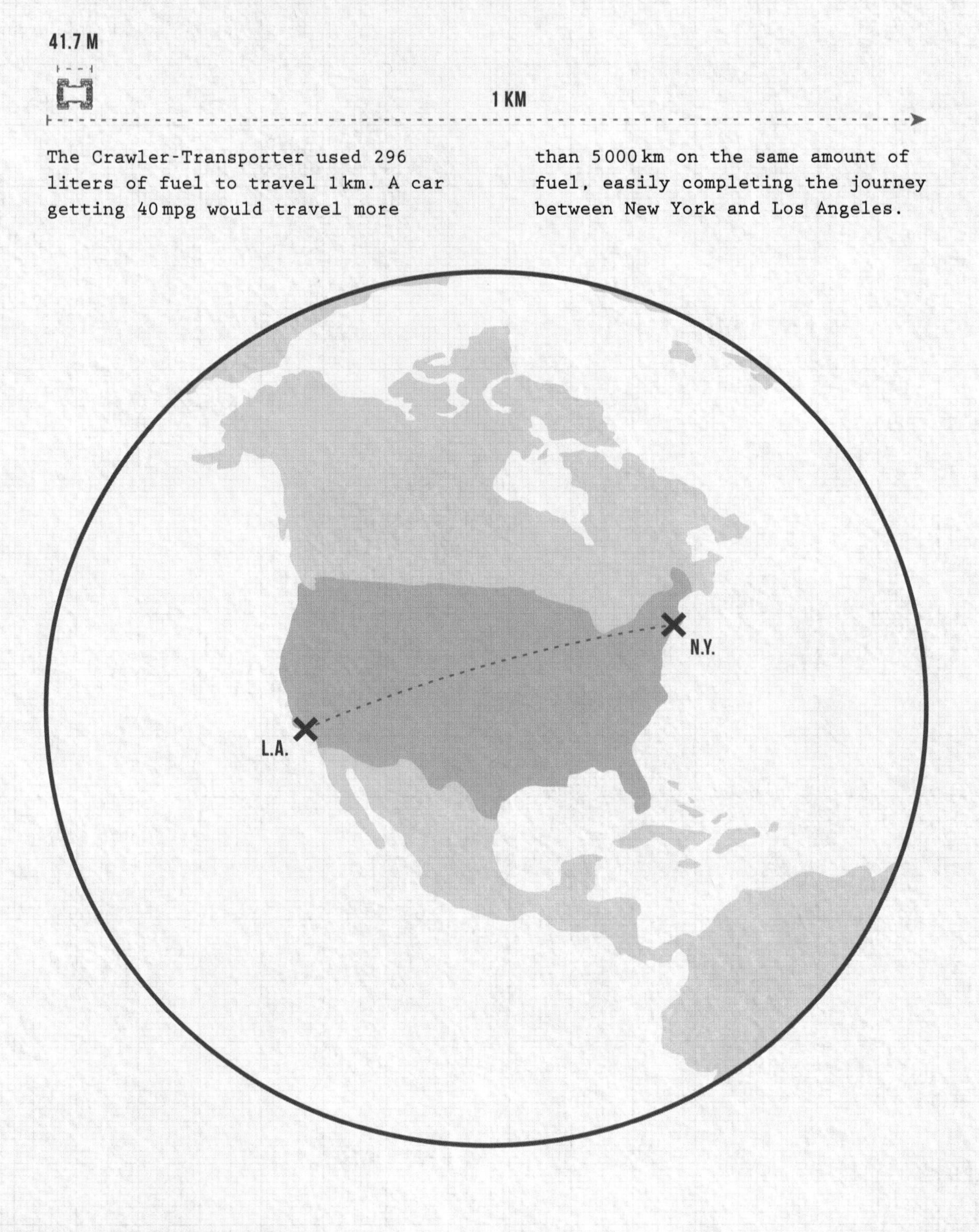

The Crawler-Transporter used 296 liters of fuel to travel 1 km. A car getting 40 mpg would travel more than 5 000 km on the same amount of fuel, easily completing the journey between New York and Los Angeles.

VEHICLE ASSEMBLY BUILDING VOLUME

The VAB's volume is the same as one and a half Great Pyramids of Giza, or the amount of air held in 1,500 hot air balloons.

VEHICLE ASSEMBLY BUILDING WEATHER

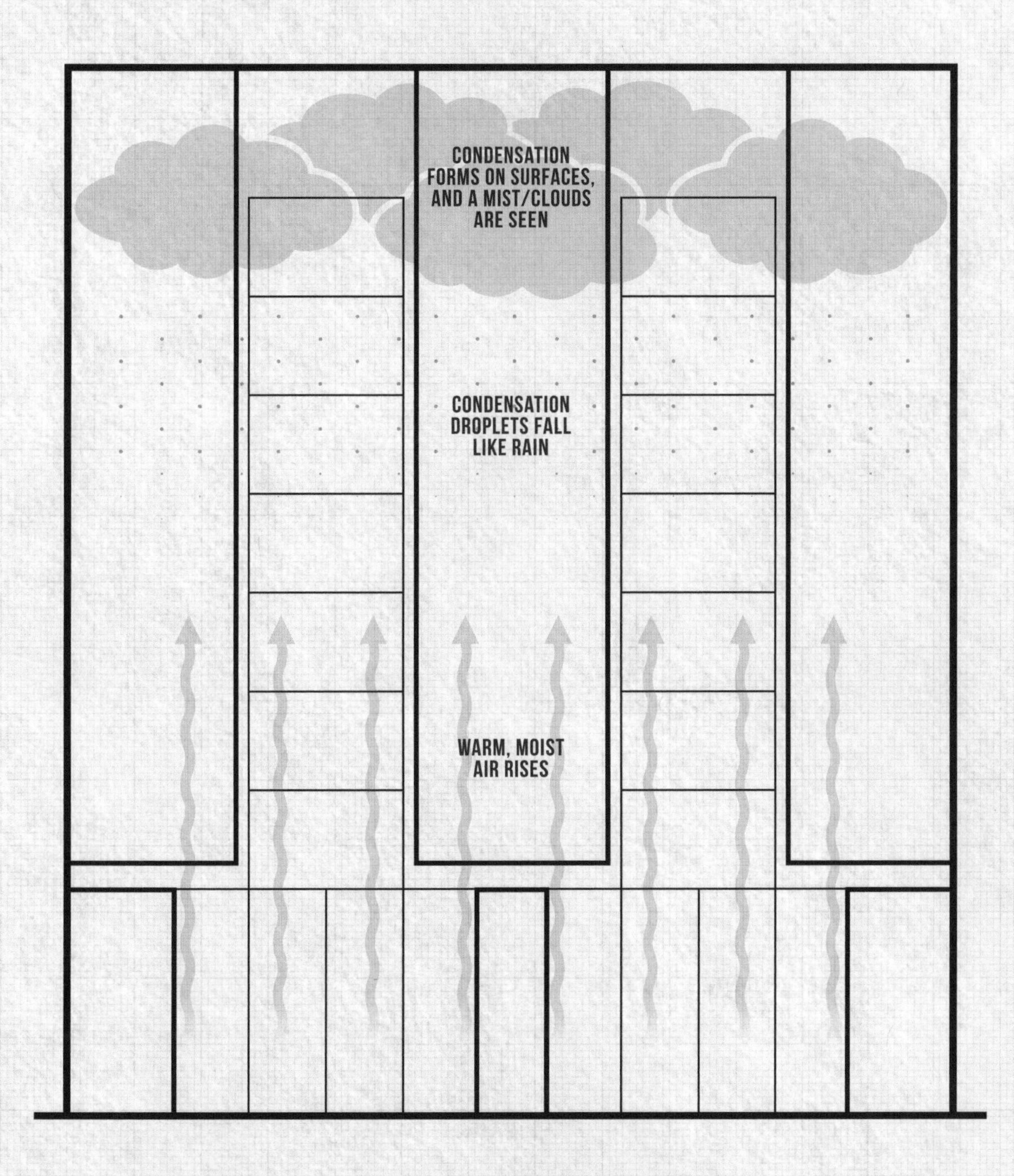

The VAB is so big that it has its own weather. Florida has a damp climate, and on especially humid days, rain clouds form beneath the ceiling.

SKYLAB – DISTANCE TRAVELLED

Skylab orbited the Earth 34,981 times, covering a distance of 140,000,000 km. This is the same as 9.36 times the distance from the Earth to the Sun.

APOLLO–SOYUZ – EARTH ORBITS

The Apollo and Soyuz spacecraft were docked together for one day, twenty-three hours, and seven minutes, during which time they covered more than 1,300,000 km, completing thirty-one orbits.

PHASES OF THE MOON

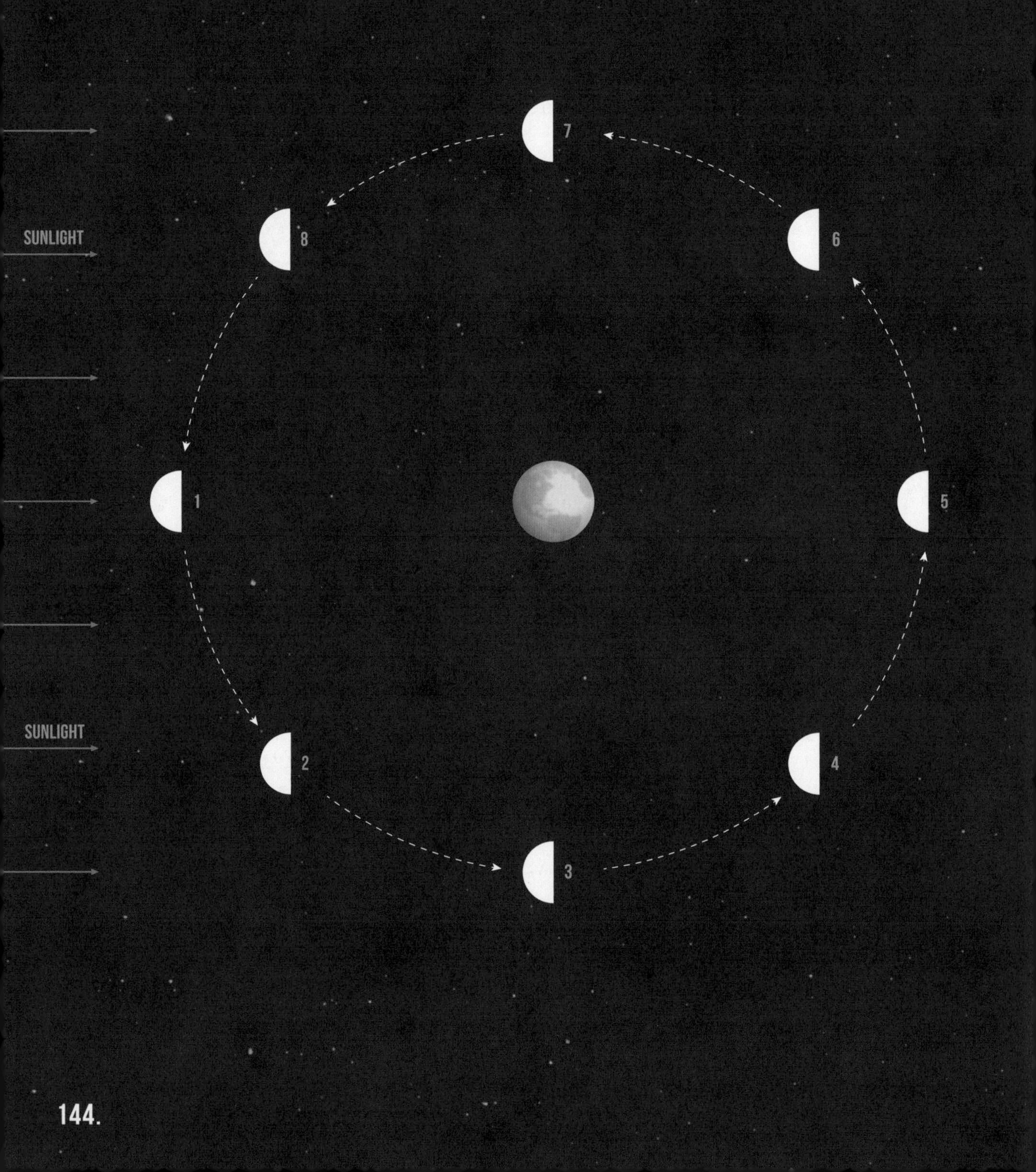

THE MOON AS SEEN FROM EARTH

LAST QUARTER
7

WANING CRESCENT
8

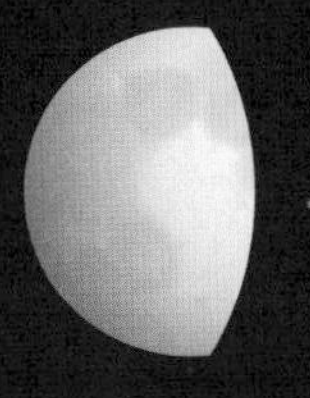

WANING GIBBOUS
6

NEW MOON
1

FULL MOON
5

WAXING CRESCENT
2

WAXING GIBBOUS
4

FIRST QUARTER
3

GLOSSARY

A7L PAGE 20

The spacesuit worn by Apollo astronauts.

ALSEP

Apollo Lunar Surface Experiments Package. Each package was a collection of scientific instruments designed to monitor the lunar environment. One was deployed near each landing site from Apollo 12 until Apollo 17.

APS PAGE 8

Ascent Propulsion System.

CAPCOM

Capsule Communicator. The individual at Mission Control Center who was responsible for communicating with the astronauts. All communications with the crew passed through a single person, and NASA felt it important that they should be another astronaut, as they would best understand the situation the crew was in and therefore provide better advice.

CM PAGES 4 AND 6

Command Module.

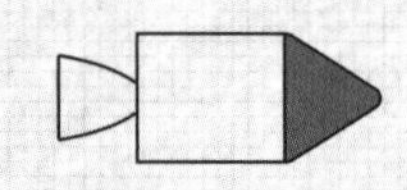

CRAWLER-TRANSPORTER PAGE 24

CSM PAGE 4 AND 6

Command/Service Module.

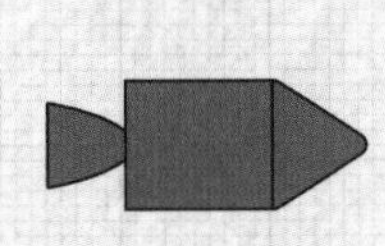

DPS PAGE 8

Descent Propulsion System.

DRAG

The force exerted on a body when it moves through the atmosphere. Caused by friction with the air, it acts in the opposite direction of travel.

DRAG

EST

Eastern Standard Time. The fifth time zone west of Greenwich. This is the time zone in which Florida lies, from where all the Apollo missions were launched.

EVA

Extra Vehicular Activity. Operations that were performed outside of the Apollo spacecraft. In orbit this was known as a spacewalk.

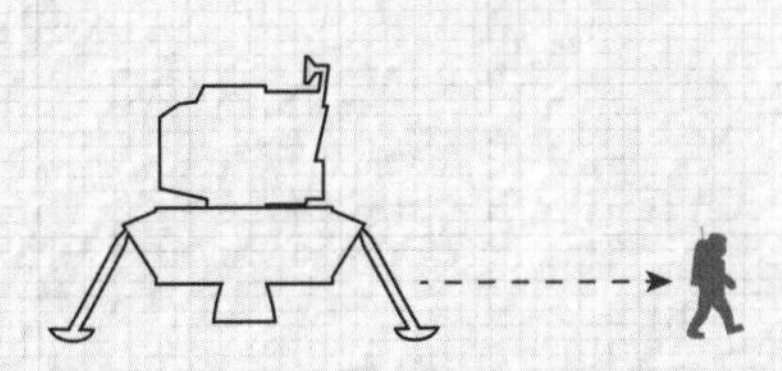

FORCE

A push or pull on an object. An object that has a force applied to it will accelerate in the direction that the force is applied. The greater the force, the greater the acceleration.

GET

Ground Elapsed Time. The amount of time that has passed since lift-off at the beginning of the mission.

GIMBAL

A device that allows a body to rotate about at least one axis. The Saturn V's F1 engines were mounted on gimbals to allow control over the direction of thrust.

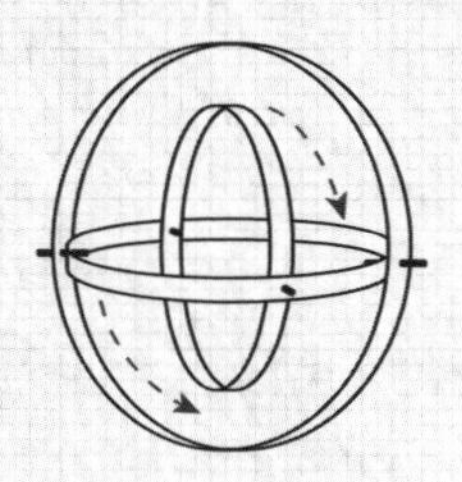

GRAVITY

The phenomenon that causes all objects in the universe to be attracted to each other. It acts over an infinite range, though its pull is lessened on objects the more distant they are.

LM PAGE 8

Lunar Module.

LRV PAGE 12

Lunar Roving Vehicle.

LUT PAGE 28

Launch Umbilical Tower.

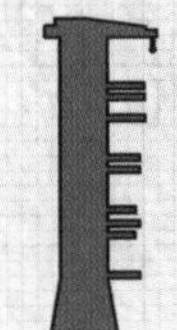

MLP PAGE 28

Mobile Launcher Platform.

NASA

National Aeronautics and Space Administration. The United States federal government agency responsible for the space program.

ORBIT

The path taken by one object around another due to gravity. These paths can be either circular or elliptical.

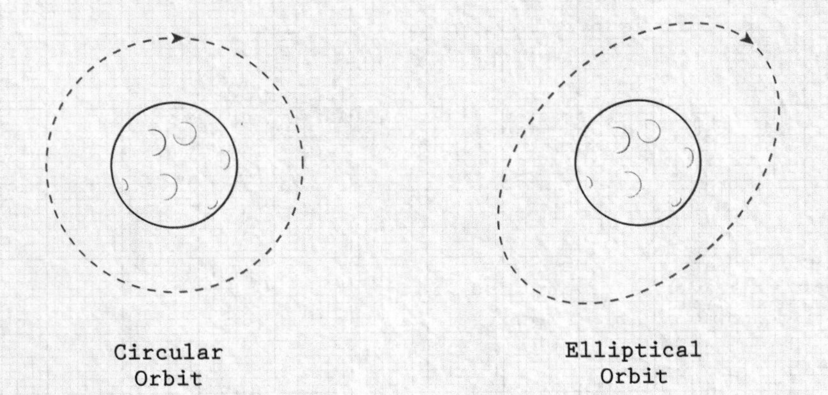

ORBITAL VELOCITY

The speed at which an object must be travelling to remain in orbit. If it is too slow it will collide with the body it is orbiting; too fast and it will break free from its orbit and float away into space. Newton's Cannonball is a thought experiment that demonstrates this.

Newton visualized a very tall mountain with a cannon at the top, and calculated the paths the cannonball would take when fired at different velocities.

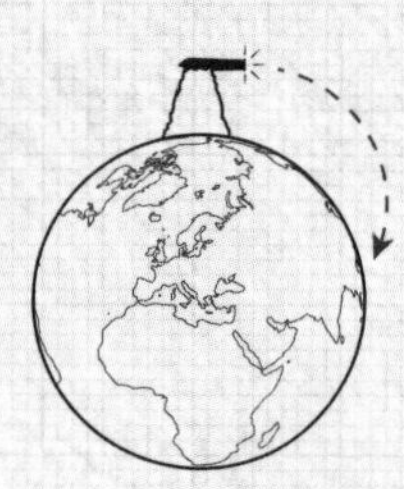

Low Velocity

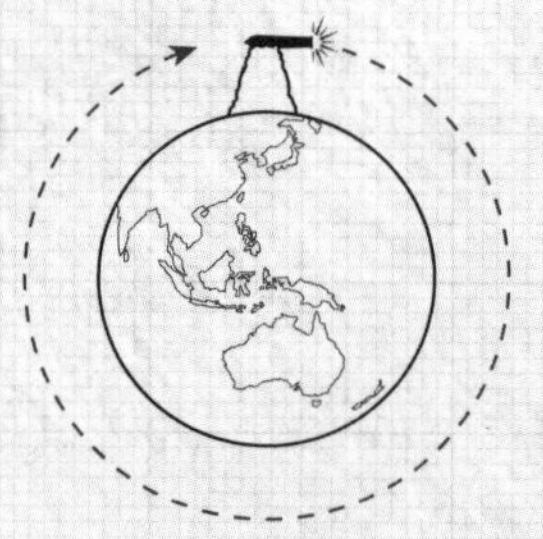

Orbital Velocity

High Velocity

PLSS PAGES 20 AND 23

Portable Life Support System.

RADAR

Radio Detection and Ranging. An object detection system that uses the information received by reflecting radio waves off objects to determine their range, velocity, or angle.

RCS PAGES 4 AND 8

Reaction Control System.

REENTRY

The moment on a mission when the Apollo CM would enter the atmosphere on returning to Earth. Friction, due to the speed it was travelling, caused huge temperatures to build up around the craft.

SATELLITE

An object that is in orbit around another. This could be man-made, such as the satellites we use to relay radio signals, or natural like the Moon.

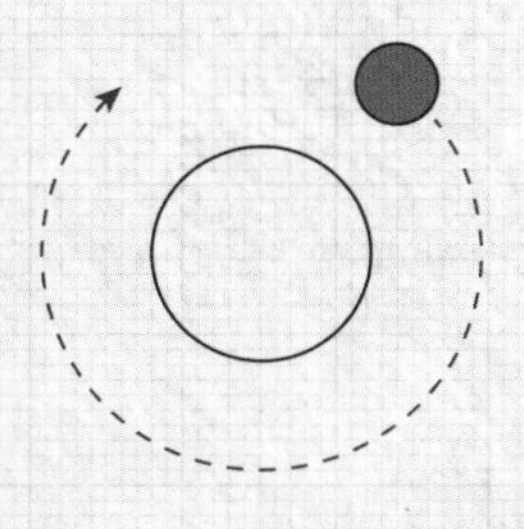

SIM

Scientific Instrument Module. Located in sector 1 of the SM from Apollo 15 onward. It housed lunar orbital sensors, cameras, film, and the subsatellite.

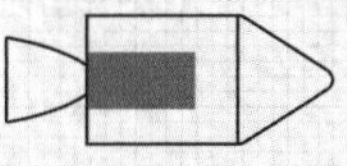

SM PAGES 4 AND 7

Service Module.

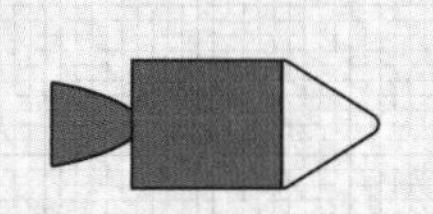

SPS PAGES 5 AND 7

Service Propulsion System.

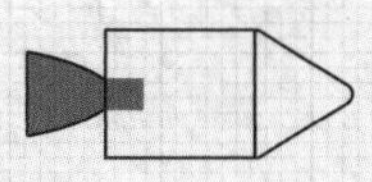

TEI

Trans-Earth Injection. The maneuver performed by the CSM to propel it from the Moon's orbit back to Earth. The SPS engine would fire, generating enough speed to send the craft on a trajectory toward Earth.

TGE

Traverse Gravimeter Experiment. This device took accurate measurements of the Moon's gravity. The findings helped scientists determine the geological makeup of the Moon.

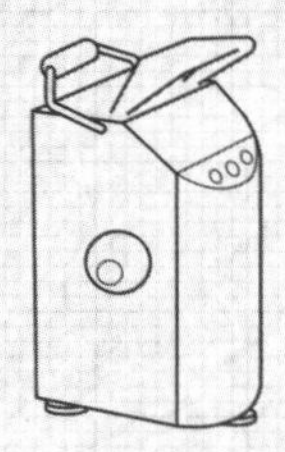

THRUST

The force that moves an aircraft or rocket through the atmosphere or space.

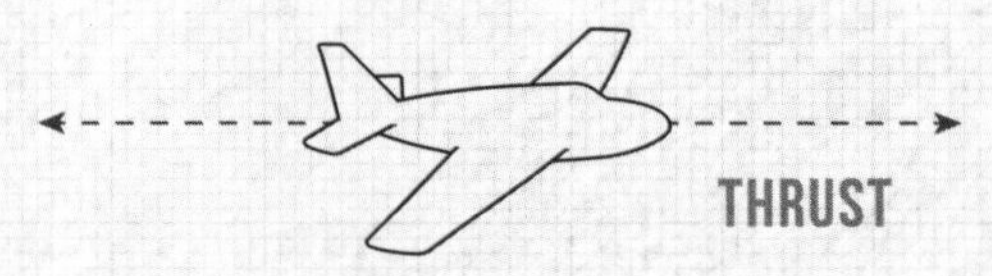

TLI

Trans-Lunar Injection. The maneuver performed by the Saturn V's third stage to fire the Apollo craft out of Earth's orbit and toward the Moon.

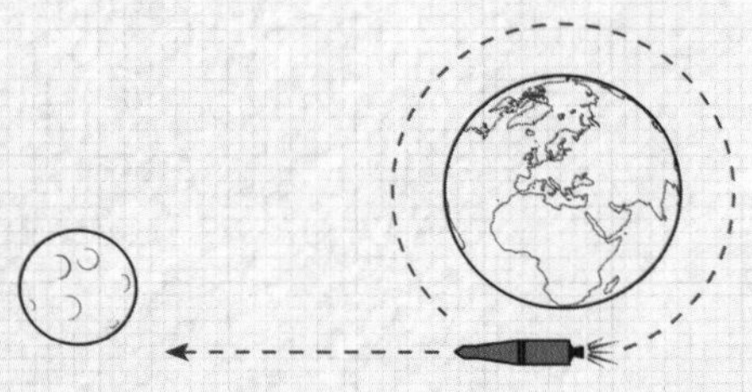

TRAJECTORY

The path followed by an object or projectile through space.

UTC

Coordinated Universal Time. The time standard used across the world, it is the same as Greenwich Mean Time (GMT) but does not observe the change to Daylight Saving Time.

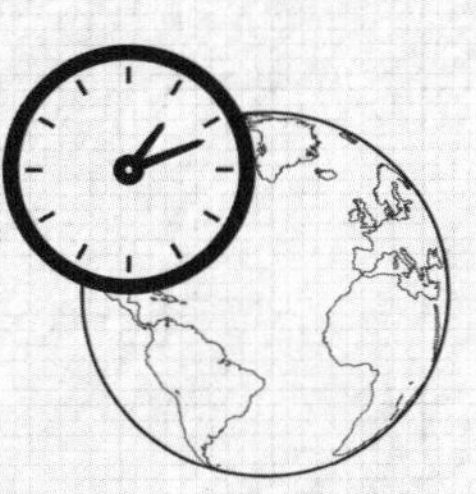

VAB P30

Vehicle Assembly Building.

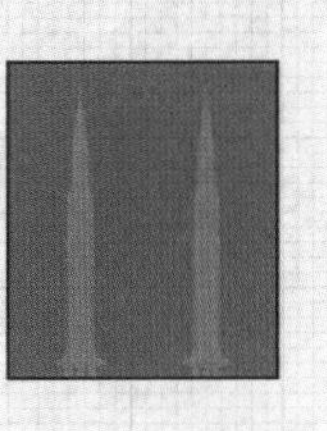

YAW, PITCH, AND ROLL

These are the three axes that aircraft and spacecraft rotate about to control the angle and direction in which they are heading.

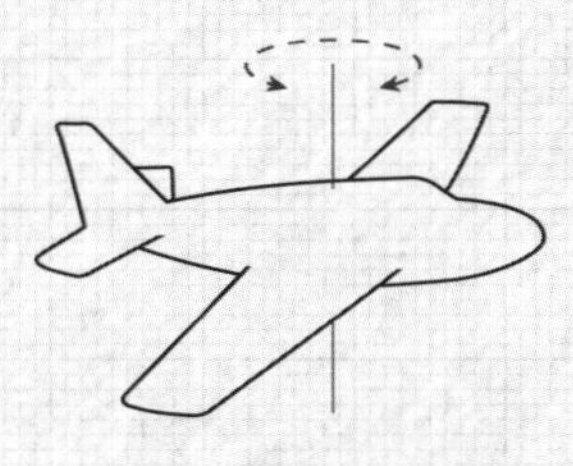

Yaw

Pitch

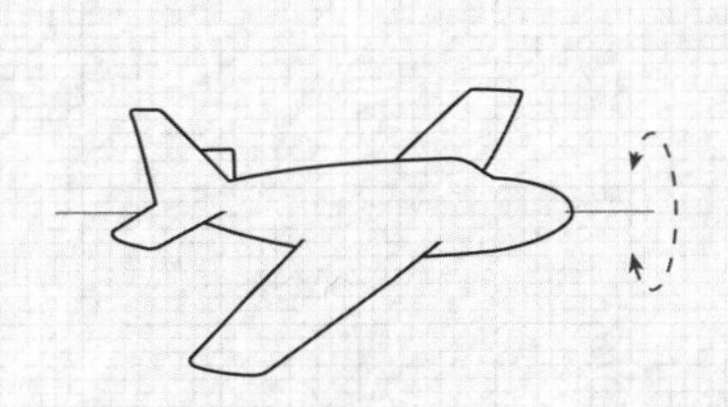

Roll

ZACHARY IAN SCOTT

BORN:
SEPTEMBER 6, 1982
CARLISLE, ENGLAND

Zack joined the Royal Air Force at the age of twenty, where he worked as an aircraft technician for several years. He then returned to civilian life to work on high-speed trains, before pursuing his lifelong passion for design. Zack achieved his degree in Graphic Design in 2013 and has since been working in-house for a couple of companies, while using his free time to work on *Apollo*, his first book. Zack has a keen interest in the sciences and loves to create graphics that make complex ideas easy to digest.

SOURCES

BOOKS

Reynolds, David West. *Apollo: The Epic Journey to the Moon, 1963-1972.* Norwalk: Easton Press, 2002.

Baker, David. *Apollo 13 Owners' Workshop Manual*. Minneapolis: Zenith Press, 2013.

Orloff, Richard W. *Apollo by the Numbers: A Statistical Reference.* Charleston: Createspace, 2013.

Harland, David M. *NASA's Moon Program: Paving the Way for Apollo 11.* New York: Springer, 2009.

Woods, W. David. *NASA Saturn V Owner's Workshop Manual*. Sparkford: Haynes Publishing, 2016.

WEBSITES

www.history.nasa.gov/apollo
Links to numerous pages that focus on the Apollo program. Broken into two sections: links that are authored by NASA, and non-NASA links.

www.nasa.gov/mission_pages
Links to summaries of all the missions, all authored by NASA.

www.airandspace.si.edu/topics/apollo-program
Displays a wide array of Apollo-related artifacts and memorabilia, including astronaut interviews, science relating to the program, and some "behind the scenes" stories.

www.apolloarchive.com
A huge archive of Apollo photography, radio transmissions, and video recordings, including links to recommended literature.

LUNAR LANDING MISSION FLIGHT PLAN

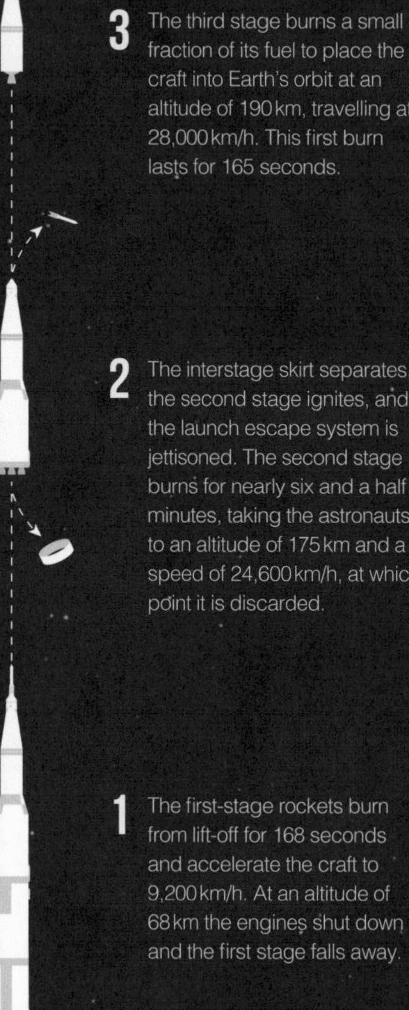

3 The third stage burns a small fraction of its fuel to place the craft into Earth's orbit at an altitude of 190 km, travelling at 28,000 km/h. This first burn lasts for 165 seconds.

2 The interstage skirt separates, the second stage ignites, and the launch escape system is jettisoned. The second stage burns for nearly six and a half minutes, taking the astronauts to an altitude of 175 km and a speed of 24,600 km/h, at which point it is discarded.

1 The first-stage rockets burn from lift-off for 168 seconds and accelerate the craft to 9,200 km/h. At an altitude of 68 km the engines shut down and the first stage falls away.

"THERE IS NO STRIFE, NO
PREJUDICE, NO NATIONAL CONFLICT
IN OUTER SPACE AS YET. ITS
HAZARDS ARE HOSTILE TO US
ALL. ITS CONQUEST DESERVES
THE BEST OF ALL MANKIND, AND
ITS OPPORTUNITY FOR PEACEFUL
COOPERATION MAY NEVER COME
AGAIN. BUT WHY, SOME SAY, THE
MOON? WHY CHOOSE THIS AS OUR
GOAL? AND THEY MAY WELL ASK,
WHY CLIMB THE HIGHEST MOUNTAIN?
WHY, THIRTY FIVE YEARS AGO,
FLY THE ATLANTIC?...

...WE CHOOSE TO GO TO THE MOON!
WE CHOOSE TO GO TO THE MOON
IN THIS DECADE AND DO THE OTHER
THINGS, NOT BECAUSE THEY ARE
EASY, BUT BECAUSE THEY ARE
HARD; BECAUSE THAT GOAL WILL
SERVE TO ORGANIZE AND MEASURE
THE BEST OF OUR ENERGIES AND
SKILLS, BECAUSE THAT CHALLENGE
IS ONE THAT WE ARE WILLING TO
ACCEPT, ONE WE ARE UNWILLING
TO POSTPONE, AND ONE WE INTEND
TO WIN."

PRESIDENT JOHN F. KENNEDY, SEPTEMBER 12, 1962

Editor: Samantha Weiner
Designer: Zack Scott
Production Manager: Mike Kaserkie

Library of Congress Control Number: 2017956771

ISBN: 978-1-4197-3219-5
eISBN: 978-1-68335-336-2

Printed and bound in USA
10 9 8 7 6 5 4 3 2 1

ABRAMS The Art of Books
195 Broadway, New York, NY 10007
abramsbooks.com